GARY H. LYMON

The Science and Business of Lab Grown Diamonds

From Creation to Building a Profitable Enterprise

optimisticdaily@gmx.us

First edition

This book was professionally typeset on Reedsy.
Find out more at reedsy.com

For the minds advancing diamond science
and the entrepreneurs building the future of the industry.
For those who refuse to separate beauty from responsibility
and choose innovation over tradition for the greater good.
This book is written for you.

Contents

Foreword

The diamond industry finds itself at a pivotal moment, balancing a rich legacy with powerful transformation. For generations, natural diamonds have fascinated the world through their brilliance, rarity, and deep cultural significance. Today, however, a new era is taking shape, guided by science, ethics, innovation, and a more conscious consumer mindset. Lab-grown diamonds are not merely challenging the industry; they are reshaping its very foundation.

In *The Science and Business of Lab Grown Diamonds*, Gary Lymon offers far more than a reference guide. He presents a thoughtful and well-balanced examination of this evolution. With clarity and confidence, he leads readers through the entire journey of lab-grown diamonds, from their scientific creation to design and customization, from certification and market adoption to their emerging role as modern heirlooms in a digital world.

This book arrives at exactly the right moment. As sustainability, transparency, and personalization become central themes across the luxury landscape, lab-grown diamonds stand out as a meaningful response to the modern demand for beauty with purpose. Gary Lymon captures this shift with both precision and genuine insight.

Whether you are a jeweler adapting to new consumer expectations, a retailer focused on building lasting trust, or a buyer seeking to understand what makes these diamonds so compelling, this guide proves to be an essential resource. It respects tradition, honors scientific progress, and champions the innovation that will shape the future of fine jewelry.

Gary H. Lymon

Gemologist, Industry Expert, Ethical Jewelry Advocate

2026

Preface

Diamonds have occupied a singular place in human history, serving as enduring symbols of love, resilience, status, and permanence. For centuries, their allure was inseparable from their origin deep within the Earth, shaped over immense spans of time and recovered through demanding processes that were often environmentally and ethically complex. Today, however, we are experiencing a meaningful transformation, one that reshapes not only how diamonds are created, but also what they signify.

The Science and Business of Lab-Grown Diamonds emerged from a growing fascination with this shift. Lab-grown diamonds represent more than a technical breakthrough; they signal a cultural and philosophical turning point. They challenge long-held assumptions about rarity and romance, and they invite a reexamination of luxury through the lens of transparency, sustainability, and intentional choice.

This book is not intended as a manifesto, but as a practical and thoughtful guide. It seeks to illuminate every dimension of the lab-grown diamond industry, from scientific foundations and market forces to consumer behavior and the emotional significance of customized jewelry. It is written for a diverse audience: jewelers navigating a changing landscape, consumers seeking responsible alternatives, students of materials science, and entrepreneurs exploring an emerging frontier.

The lab-grown diamond sector is evolving at a remarkable pace. In writing this book, my aim was not only to document that evolution, but also to provide a framework for understanding where the industry has been, where it stands today, and where it may be headed.

Whether you are reading this in a retail showroom, a design studio, or at home, I hope it offers clarity, perspective, and inspiration around one of the most compelling transformations taking place in the jewelry world today.

-Gary Lymon

Acknowledgments

Writing *The Science and Business of Lab-Grown Diamonds* has been a journey shaped by discovery, collaboration, and genuine appreciation. Along the way, I have been fortunate to learn from and be supported by many individuals whose contributions made this book possible.

To the scientists, gemologists, and engineers who continue to expand the boundaries of what is possible, your work forms the very foundation of this story. Thank you for your generosity in sharing knowledge and for demonstrating that innovation and ethical responsibility can advance hand in hand.

To the retailers, designers, and entrepreneurs redefining the jewelry industry from within, your insight and vision gave this book its practical relevance and real-world perspective.

To my editors, researchers, and early readers, your thoughtful feedback, encouragement, and careful attention helped shape every chapter into something that truly reflects the importance of this subject.

To my family and friends, your constant support, curiosity, and belief in the value of this work sustained me throughout the process and kept me moving forward.

And finally, to the readers, whether you approach this book through professional interest or personal curiosity, thank you for seeking understanding in a rapidly evolving field. I hope this guide provides you with both clarity and inspiration as you explore the world of lab-grown diamonds.

Gary H. Lymon

Chapter 1

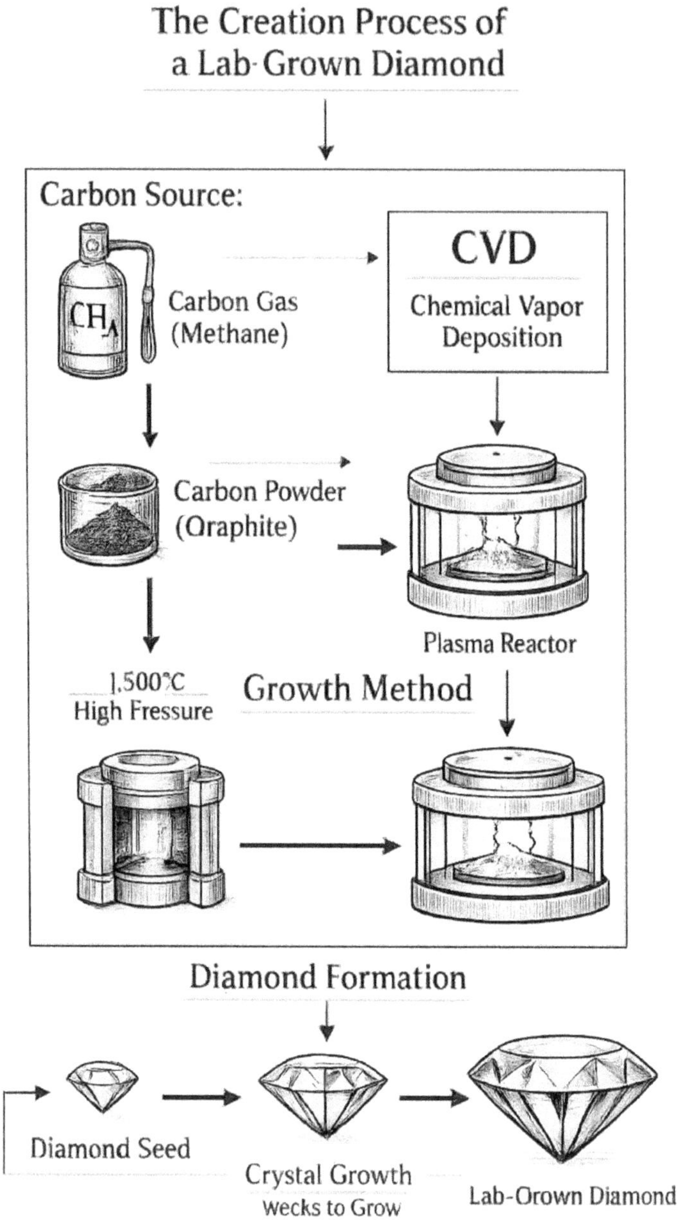
The Creation Process of a Lab-Grown Diamond
Carbon Source:
CH_4
Carbon Gas (Methane)
CVD
Chemical Vapor Deposition
Carbon Powder (Oraphite)
Plasma Reactor
1,500°C
High Fressure
Growth Method
Diamond Formation
Diamond Seed
Crystal Growth
Wecks to Grow
Lab-Orown Diamond

Introduction to Lab-Grown Diamonds

In a world increasingly shaped by rapid technological progress, heightened environmental awareness, and evolving consumer priorities, lab-grown diamonds have emerged as one of the most transformative developments in the modern jewelry industry. For generations, diamonds have stood as enduring symbols of luxury, exclusivity, and timeless elegance. They have long been woven into human history, marking life's most meaningful moments through engagement rings, fine jewelry, and treasured family heirlooms passed down through time.

Yet, beneath this long-standing tradition, a quiet but powerful transformation has been taking place. Over the last few decades, lab-grown diamonds have moved from experimental creations confined to scientific research facilities into widely accepted and sought-after gemstones. Once considered a fringe concept, they have steadily gained legitimacy, reshaping both the diamond trade and the broader luxury market. Their rise represents not simply a technological achievement, but a shift in how value, ethics, and beauty are defined in the modern era.

This chapter introduces the dynamic world of lab-grown diamonds, examining their origins, their increasing presence in the global market, and the forces fueling their rapid growth. It explores the fundamental distinctions between mined and lab-grown diamonds, the innovations that made laboratory production possible, and the changing consumer values that have accelerated this transformation. Together, these elements reveal how the diamond industry is undergoing one of the most significant evolutions in its history.

The Diamond Industry: Natural Diamonds and Lab-Grown Alternatives

For centuries, the diamond industry has been built around the mystique of natural diamonds, formed deep beneath the Earth's surface over millions of years. These gemstones have traditionally been valued for their rarity, natural origin, and association with life's most meaningful milestones. However, as demand has increased, so too has scrutiny of the processes required to extract these stones from the ground.

Natural diamonds are recovered through methods such as open-pit mining, underground mining, and marine mining. Each of these approaches is highly resource-intensive, involving vast amounts of land disruption, energy consumption, labor, and financial investment. The path from rough stone to polished gem is complex and costly, often leaving a substantial environmental footprint. Issues such as deforestation, ecosystem destruction, and carbon emissions have become central concerns in discussions about the long-term sustainability of diamond mining.

Lab-grown diamonds present a fundamentally different approach. Also referred to as synthetic or cultured diamonds, they are produced in controlled laboratory settings using two primary techniques: High Pressure High Temperature (HPHT) and Chemical Vapor Deposition (CVD). These methods recreate the natural conditions required for diamond formation, but they do so over a dramatically shorter period, typically weeks or months rather than millions of years.

Unlike traditional mining, lab-grown diamond production does not involve large-scale excavation or environmental disruption. This distinction has positioned lab-grown diamonds as a compelling alternative for consumers who are increasingly attentive to ecological responsibility and ethical sourcing. Though their origin differs, lab-grown diamonds are identical to mined diamonds in their chemical composition, physical structure, and optical performance.

Sustainability, Ethics, and Changing Consumer Preferences

Over recent decades, consumer behavior has shifted in meaningful ways. Buyers today are more informed, more values-driven, and more conscious of the broader impact of their purchasing decisions. This shift has been especially visible in the luxury sector, where sustainability, transparency, and ethical accountability are now central considerations rather than secondary concerns.

The diamond industry has not been immune to these changes. Growing awareness of environmental damage and human rights issues linked to traditional diamond mining has prompted many consumers to seek alternatives. One of the most influential factors in this shift has been global attention on conflict diamonds, often referred to as blood diamonds. These stones, mined in regions affected by armed conflict and sold to finance violence, have exposed serious ethical flaws within parts of the diamond supply chain.

Lab-grown diamonds offer a clear response to these concerns. Because they are created in controlled environments with fully traceable origins, they eliminate the risks associated with conflict financing and exploitative labor practices. For consumers who wish to enjoy the beauty and symbolism of diamonds without ethical compromise, lab-grown options provide a compelling solution.

Affordability has also played a significant role in their growing popularity. Natural diamonds have long been associated with exclusivity, but rising prices and limited supply have placed them beyond the reach of many buyers. Lab-grown diamonds, by contrast, are typically priced lower than mined diamonds of comparable quality. This has broadened access to diamond ownership, allowing a wider range of consumers to participate in what was once a highly exclusive market.

Technology and the Transformation of the Diamond Market

The emergence of lab-grown diamonds is rooted in extraordinary advances in science and engineering. The development of HPHT and CVD technologies represents a major breakthrough in materials science, enabling the creation of diamonds that are virtually indistinguishable from their natural counterparts.

These processes replicate the conditions under which diamonds naturally form, producing stones with identical chemical, physical, and optical properties. As a result, lab-grown diamonds exhibit the same brilliance, fire, hardness, and durability as mined diamonds. To the naked eye and even under professional examination, the differences are undetectable without advanced testing.

Technological progress has also expanded the possibilities of diamond production. Lab-grown diamonds can now be created in a wide range of colors, shapes, and sizes, offering consumers greater choice and customization than ever before. Advances in cutting and polishing techniques have further enhanced quality, allowing lab-grown diamonds to achieve exceptional clarity and precision that rivals the finest natural stones.

Looking Ahead: The Future of the Diamond Industry

As the diamond market continues to evolve, lab-grown diamonds are positioned to play an increasingly influential role. While natural diamonds are likely to retain their cultural and historical significance, the rapid expansion of the lab-grown sector reflects a fundamental change in how luxury is perceived and valued.

In the chapters that follow, this book will explore the science behind lab-grown diamonds, the technologies used to create them, their applications across industries, and their environmental and economic impact. It will also address common misconceptions, provide guidance for selecting lab-grown diamonds, and examine how this growing industry is reshaping the global market.

Whether you are a consumer considering a purchase, a professional navigating a changing industry, or simply curious about the future of diamonds, this guide offers a comprehensive and accessible exploration of lab-grown diamonds. By the end of this book, you will have a clear understanding of their benefits, their role in modern luxury, and the broader shift they represent.

The rise of lab-grown diamonds is more than a trend. It signals a deeper transformation in how beauty, responsibility, and value are defined. Welcome to the future of diamonds.

Chapter 1: What Is a Lab-Grown Diamond?

When most people imagine a diamond, they picture a rare gemstone formed deep within the Earth, created over millions of years and unearthed through intensive mining. This traditional image has long defined what a diamond is supposed to be: ancient, scarce, and naturally occurring. However, the emergence of lab-grown diamonds has challenged that perception, introducing a new category of diamonds that stands firmly alongside mined stones in the modern jewelry landscape.

This raises an important question: what exactly is a lab-grown diamond, and how does it differ from a natural one? In this chapter, we will clearly define lab-grown diamonds, explain the scientific principles behind their creation, and explore the characteristics that make them virtually indistinguishable from mined diamonds. We will also examine the key differences between natural and lab-grown diamonds, focusing on their origins, properties, and how they are perceived by consumers.

Definition and Core Principles of Lab-Grown Diamonds

A lab-grown diamond is a genuine diamond, not a substitute or imitation. It is composed entirely of carbon atoms arranged in the same crystal structure found in natural diamonds. The defining difference lies not in what the diamond is, but in how and where it is formed.

Lab-grown diamonds are created in laboratory environments using advanced technologies that replicate the natural conditions under which diamonds develop deep within the Earth's mantle. The two primary methods used are High Pressure High Temperature (HPHT) and Chemical Vapor Deposition (CVD). Both techniques recreate the intense heat and pressure required for diamond formation, but they do so in a controlled setting and over a much shorter time span.

Despite their laboratory origin, lab-grown diamonds share the same chemical composition, crystal lattice, and physical properties as mined diamonds. The key distinction is time. While natural diamonds require millions of years to form, lab-grown diamonds can be produced in a matter of weeks. This efficiency does not compromise quality; instead, it allows for precision and consistency that is difficult to achieve in nature.

Like natural diamonds, lab-grown diamonds are available in a wide variety of shapes, sizes, and colors. They can be cut and polished using the same techniques and are recognized as real diamonds by gemologists and jewelers worldwide. Their legitimacy within the diamond industry is firmly established.

Comparing Natural Diamonds and Lab-Grown Diamonds

Although lab-grown and natural diamonds are identical in composition and performance, there are important distinctions between them, particularly in terms of origin, cost, and environmental impact.

- **Origin**Natural diamonds form approximately 100 miles beneath the Earth's surface under extreme heat and pressure. Over millions of years, carbon atoms bond into a rigid crystal structure, eventually creating a diamond. Lab-grown diamonds follow the same fundamental process, but it takes place in a laboratory using HPHT or CVD technology, where conditions are carefully controlled to replicate nature.
- **Physical and Chemical Properties**From a scientific perspective, lab-grown and natural diamonds are the same. Both consist of pure carbon arranged in a crystal lattice that gives diamonds their extraordinary hardness, brilliance, and durability. Lab-grown diamonds rank a 10 on the Mohs hardness scale, just like mined diamonds, and exhibit the same light performance, thermal conductivity, and optical characteristics. Even trained professionals often cannot tell them apart without specialized

equipment.

- **Cost**One of the most noticeable differences between the two lies in price. Natural diamonds can be costly due to mining expenses, limited supply, and market dynamics related to rarity. Lab-grown diamonds are generally more affordable because they eliminate the need for extensive extraction, transportation, and large-scale mining infrastructure. As production technology continues to improve, lab-grown diamonds are becoming even more cost-effective.
- **Rarity and Availability**Natural diamonds are considered rare because of the immense time and specific geological conditions required for their formation. This rarity contributes to their perceived value and mystique. Lab-grown diamonds, by contrast, are not inherently scarce. They can be produced consistently as long as the necessary technology and resources are available, making them a more accessible option for consumers.
- **Environmental and Ethical Considerations**Diamond mining has long raised concerns related to environmental damage, including deforestation, water pollution, and habitat destruction. It is also associated with ethical issues such as unsafe labor conditions and conflict diamonds. Lab-grown diamonds avoid these challenges by eliminating mining altogether. Produced in transparent, regulated environments, they are widely viewed as a more ethical and environmentally responsible choice.

Chemical, Physical, and Optical Characteristics

To fully understand why lab-grown diamonds are considered true diamonds, it is important to examine their fundamental properties.

- **Chemical Composition**Both lab-grown and natural diamonds consist entirely of carbon. The carbon atoms are bonded in a tetrahedral lattice structure, which gives diamonds their remarkable strength and resistance to wear. This identical chemical makeup ensures that lab-grown diamonds perform exactly like mined diamonds in everyday use.
- **Physical Properties**Diamonds are the hardest known natural material, a

quality that makes them exceptionally durable. Lab-grown diamonds share this same hardness and structural integrity. They also exhibit similar thermal conductivity, which is why diamonds, both natural and lab-grown, are used in industrial applications such as cutting tools and heat management systems.

- **Optical Performance**A diamond's beauty lies in how it interacts with light. Brilliance, fire, and scintillation are all products of the diamond's internal structure and precise cutting. Because lab-grown diamonds have the same crystal lattice as natural diamonds, they reflect and refract light in exactly the same way, producing the same dazzling visual effect.
- **Color**Diamonds naturally occur in a range of colors, from completely colorless to yellow, brown, and rare hues such as blue, pink, and green. Color is graded on a scale from D, which is colorless, to Z, which indicates visible color. Lab-grown diamonds can be produced across this same spectrum. Through careful control of the growth environment, specific colors can be created intentionally, offering consumers a wider selection at often lower prices than naturally colored diamonds.

Why Lab-Grown Diamonds Continue to Gain Popularity

The growing acceptance of lab-grown diamonds is driven by a combination of factors that align closely with modern consumer values.

- **Sustainability**Lab-grown diamonds have a significantly smaller environmental footprint than mined diamonds. As concerns about climate change and environmental preservation increase, many consumers are choosing options that minimize ecological harm.
- **Ethical Responsibility**Lab-grown diamonds eliminate concerns related to conflict financing, unsafe labor practices, and lack of transparency. For buyers who value ethical sourcing, this provides peace of mind without sacrificing beauty or quality.
- **Affordability**By removing mining and reducing production costs, lab-grown diamonds offer exceptional value. Consumers can often choose

larger or higher-quality stones for the same budget, making diamonds more accessible than ever before.

In the next chapter, we will explore the history of lab-grown diamonds, tracing their journey from early scientific experimentation to their current position as a rapidly expanding and influential industry.

Chapter 2: The History of Lab-Grown Diamonds

The history of lab-grown diamonds is a story driven by curiosity, determination, and scientific ambition. What began as an effort to understand one of nature's most extraordinary materials has evolved into a global industry that challenges long-held beliefs about how diamonds are created and valued. Tracing this history offers more than a timeline of technological progress. It reveals how shifting attitudes toward sustainability, ethics, and consumer choice have reshaped the jewelry industry itself.

This chapter explores the origins of lab-grown diamonds, beginning with early scientific experimentation and moving through the breakthroughs that made commercial production possible. Along the way, we will examine the individuals and innovations that propelled the technology forward and the key milestones that established lab-grown diamonds as a credible and respected alternative to mined stones.

Early Research and Breakthroughs in Diamond Synthesis

The idea of creating diamonds in a laboratory is far older than many people realize. Although commercial success is relatively recent, scientific interest in synthetic diamonds dates back more than a century. Researchers have long been fascinated by diamonds because of their exceptional physical properties and their simple chemical composition: pure carbon.

During the nineteenth century, scientists began exploring whether it might be possible to recreate the extreme pressure and temperature conditions responsible for natural diamond formation. Early experiments attempted to compress carbon under high pressure, but the technology of the time was insufficient to achieve meaningful results. While these early efforts were largely unsuccessful, they laid the conceptual foundation for future breakthroughs.

True progress did not arrive until the mid-twentieth century. In 1954, a major milestone was achieved when researchers at the General Electric Company in the United States successfully synthesized diamonds under laboratory conditions. The team, led by scientist Tracy Hall, managed to replicate the intense heat and pressure found deep within the Earth's mantle using what would later become known as the High-Pressure High Temperature method.

The result was a tiny synthetic diamond measuring approximately one millimeter. Though small and unsuitable for jewelry, this achievement proved that diamonds could be created artificially. For the first time, diamond synthesis moved from theory into reality, marking a turning point in materials science.

The Pioneers Who Advanced Lab-Grown Diamonds

Tracy Hall's breakthrough marked the beginning of synthetic diamond production, but it did not immediately lead to widespread commercial use. In the decades that followed, researchers and engineers focused on refining the process, improving diamond quality, and identifying practical applications.

During the 1970s and 1980s, improvements in HPHT technology made it possible to produce larger and more consistent diamonds. These stones were primarily used for industrial purposes rather than jewelry. Their exceptional hardness made them ideal for cutting tools, abrasives, and wear-resistant materials used in manufacturing, construction, and engineering.

As demand for industrial diamonds grew, synthetic production offered a reliable and scalable solution. Manufacturers could produce diamonds with specific characteristics, reducing dependence on natural diamond supplies. This period established lab-grown diamonds as a valuable industrial resource, even though their role in fine jewelry was still limited.

Key Milestones in the Development of Lab-Grown Diamonds

The modern lab-grown diamond industry began to take shape as technological advances aligned with changing market demands and consumer values.

The 1980s: Industrial Expansion and Early Commercial Use

By the 1980s, HPHT-produced diamonds were widely used in industrial settings. While these early synthetic diamonds often lacked the clarity and color needed for jewelry, they played a crucial role in manufacturing and heavy industry. Controlled production allowed companies to meet growing demand without relying solely on mined diamonds, reducing pressure on natural resources.

The 1990s: Technological Progress and Early Jewelry Potential

The 1990s introduced a new chapter with the advancement of Chemical Vapor Deposition technology. Unlike HPHT, the CVD process involved breaking down carbon-rich gases in a vacuum chamber and depositing carbon atoms onto a substrate to form diamond crystals. This approach offered greater control over purity and crystal structure.

As CVD technology improved, the quality of lab-grown diamonds began to approach that of natural diamonds. Although still limited in availability, these diamonds showed promise for use in fine jewelry. The groundwork was being laid for a broader commercial shift.

The 2000s: Entry into the Consumer Market

The early 2000s marked a decisive turning point. Advances in both HPHT and CVD methods enabled the production of diamonds suitable for jewelry in a wide range of sizes, colors, and shapes. Companies such as Gemesis, later known as Pure Grown Diamonds, and Mouawad began offering lab-grown diamonds to consumers.

These diamonds were chemically identical to mined diamonds but available at lower prices. At the same time, growing awareness of sustainability and ethical sourcing issues helped lab-grown diamonds gain acceptance in the luxury market. What was once experimental technology was becoming a commercial reality.

The 2010s: Rapid Growth and Mainstream Acceptance

The 2010s saw explosive growth in both production capacity and consumer adoption. As costs declined and quality improved, lab-grown diamonds became accessible to a much wider audience. Jewelry brands began incorporating them into their collections, offering consumers more choice than ever before.

Even traditional industry leaders began to adapt. De Beers, once a vocal opponent of lab-grown diamonds, entered the market through its Lightbox Jewelry brand. This move signaled a major shift in industry attitudes and further legitimized lab-grown diamonds in the eyes of consumers.

Greater transparency around sourcing, grading, and production also contributed to rising consumer confidence. Buyers increasingly valued the ability to trace their diamonds and understand how they were made.

The 2020s: Industry Maturity and Market Confidence

By the 2020s, lab-grown diamonds were firmly established within the global jewelry market. Industry forecasts suggested strong continued growth, with lab-grown diamonds expected to claim a significant share of total diamond sales. Consumer demand for ethical, sustainable, and responsibly produced luxury goods showed no signs of slowing.

Lab-grown diamonds were no longer viewed as alternatives of last resort. They had become a preferred choice for many buyers who valued transparency, innovation, and value alongside beauty.

Technology as the Driving Force

Advances in materials science have been central to the success of lab-grown diamonds. Early synthetic diamonds were small and flawed, but continuous research transformed experimental techniques into reliable production methods.

The HPHT process remains widely used, particularly for producing diamonds with specific structural characteristics. Meanwhile, the CVD method has gained prominence for its precision and ability to produce diamonds with exceptional clarity and consistency. Ongoing experimentation continues to refine both techniques, improving efficiency and reducing costs.

These technological advancements have expanded the role of lab-grown diamonds beyond jewelry into electronics, medicine, and advanced manufacturing, further strengthening the industry's long-term potential.

Looking Forward

The history of lab-grown diamonds reflects a broader story of innovation and disruption. From their origins as tiny industrial stones to their current status as luxury gemstones, lab-grown diamonds have reshaped the diamond industry. Their rise mirrors growing consumer demand for ethical sourcing, environmental responsibility, and value-driven luxury.

In the next chapter, we will take a closer look at the science behind lab-grown diamond creation. We will examine how the High-Pressure High Temperature and Chemical Vapor Deposition processes work, how they differ, and how each method influences the characteristics of the final diamond.

Chapter 3: The Technology Behind Lab-Grown Diamonds

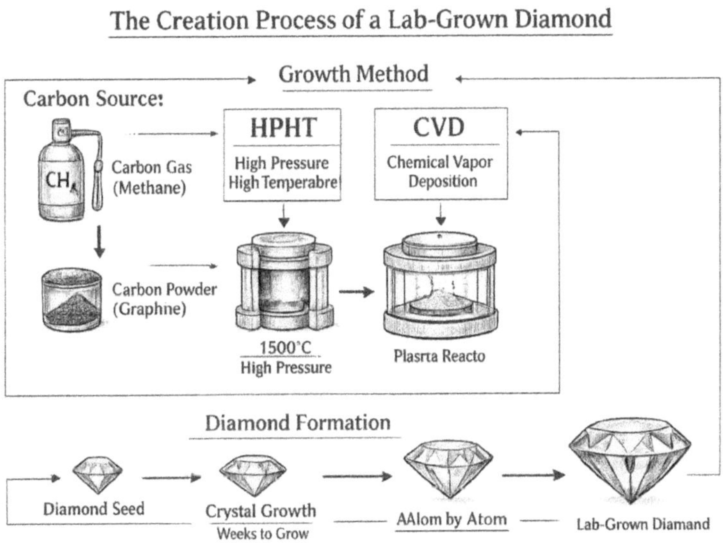

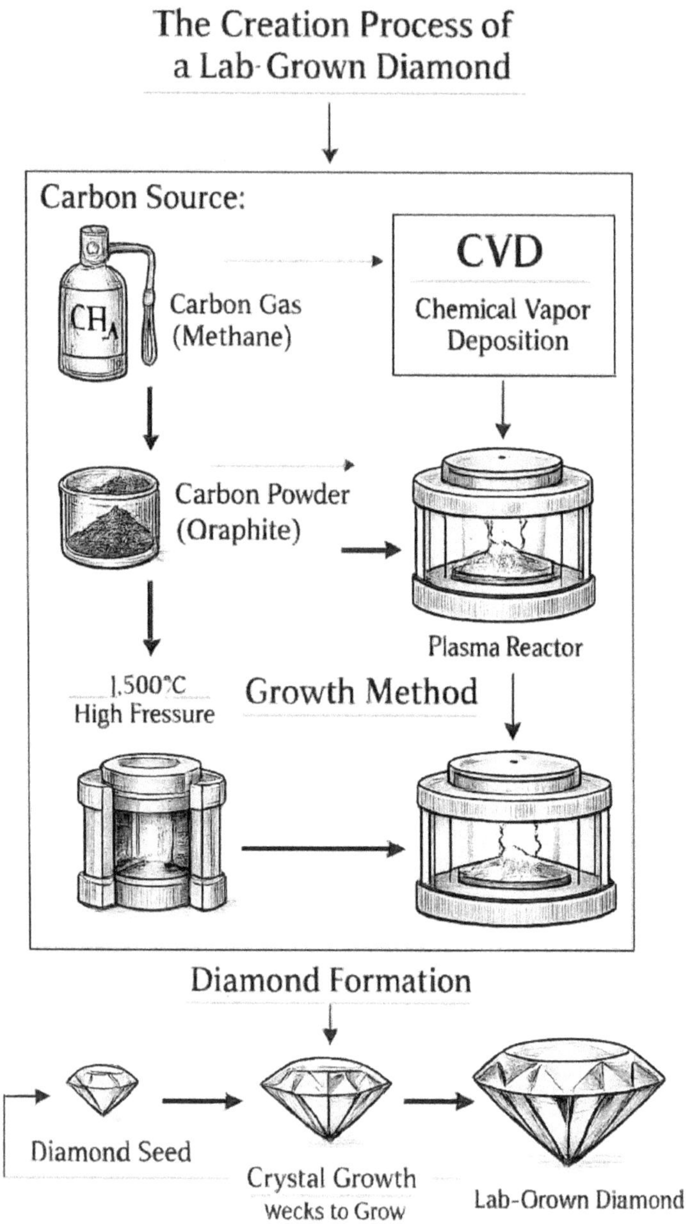
The Creation Process of a Lab-Grown Diamond
Carbon Source:
CVD
Chemical Vapor Deposition
CH_4
Carbon Gas (Methane)
Carbon Powder (Oraphite)
Plasma Reactor
1,500°C High Fressure
Growth Method
Diamond Formation
Diamond Seed
Crystal Growth
Wecks to Grow
Lab-Orown Diamond

The process of creating diamonds in a laboratory setting is nothing short of extraordinary. Diamonds, formed deep within the Earth over millions of years, are now being replicated in a controlled environment thanks to human ingenuity and technological advancement. The methods used to create lab-grown diamonds have evolved significantly over the last few decades, resulting in stones that are indistinguishable from their mined counterparts both in appearance and in chemical composition.

Understanding the science behind lab-grown diamonds not only deepens our appreciation of the technology but also highlights the ways in which we've pushed the boundaries of material science to replicate one of nature's most precious creations.

In this chapter, we will explore the two main methods used to create lab-grown diamonds: the High Pressure, High Temperature (HPHT) method and the Chemical Vapor Deposition (CVD) method. We will examine how each process works, the advantages and limitations of each, and how they have evolved over time to meet the increasing demand for high-quality, sustainable diamonds.

High Pressure, High Temperature (HPHT) Process: The First Method

The High Pressure, High Temperature (HPHT) method is the original technique for creating lab-grown diamonds. First developed in the 1950s, this process mimics the natural conditions under which diamonds are formed deep within the Earth's mantle, where temperatures can reach 1,300°C to 1,600°C (2,400°F to 2,900°F) and pressures can exceed 5 gigapascals (about 50,000 times the pressure of Earth's atmosphere).

The HPHT process begins with a small "seed" diamond, typically a tiny fragment of natural diamond. This seed is placed inside a press that generates the extreme heat and pressure needed for diamond formation. Carbon (usually in the form of graphite) is then added, and under the intense heat and pressure, the carbon atoms bond in a crystalline structure, forming a diamond.

Steps in the HPHT Process

1. **Seed Placement**: A small diamond seed is placed inside a metal chamber

along with a carbon source, usually graphite.

2. **Application of Pressure and Heat**: The chamber is subjected to extreme pressure and heat, replicating conditions found deep within the Earth.
3. **Formation of Diamond Crystals**: Under these conditions, the carbon atoms bond together in the crystal lattice structure that defines a diamond.
4. **Growth Process**: The diamond crystal grows layer by layer over time, forming a larger, more perfect diamond.

Diagram of the HPHT Process

Here's a simple illustration of how the HPHT process works:

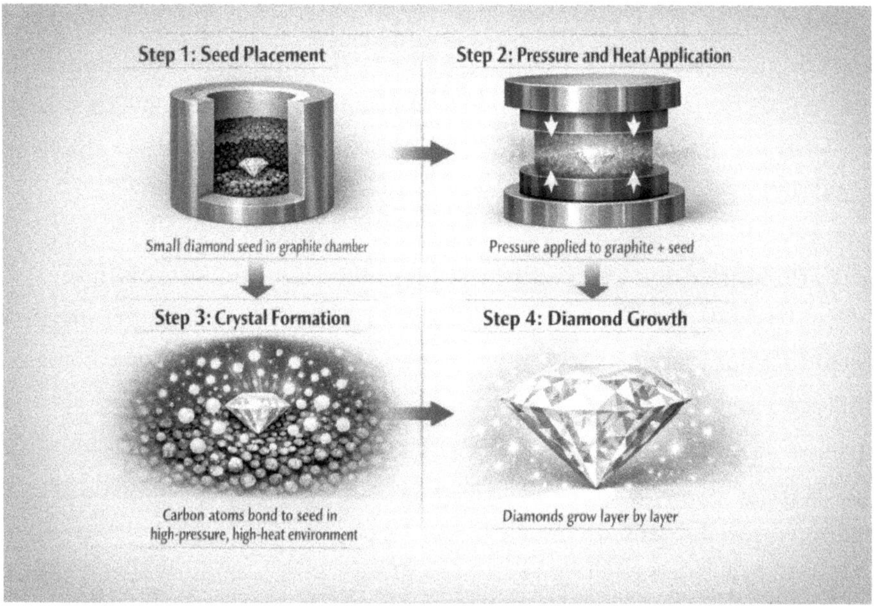

Advantages of HPHT:

- **Replicates Natural Conditions**: Since this method replicates the natural process, it can create diamonds that are structurally and chemically identical to mined diamonds.

- **Speed**: HPHT diamonds can be created relatively quickly compared to natural diamonds, although they still require a significant amount of time.

Limitations of HPHT:

- **Size Limitations**: While HPHT can create high-quality diamonds, the process is better suited for producing smaller diamonds, typically under 2 carats.
- **Energy Consumption**: The high temperatures and pressures required for this method consume significant amounts of energy.

Chemical Vapor Deposition (CVD) Method: A New Approach

The Chemical Vapor Deposition (CVD) method was developed in the 1980s as an alternative to HPHT. CVD has become increasingly popular due to its ability to produce larger diamonds and its flexibility in terms of the types of diamonds that can be grown. Unlike HPHT, which relies on extreme heat and pressure, CVD uses a process that involves breaking down a gas mixture (usually methane and hydrogen) in a vacuum chamber, where the carbon atoms from the gas bond together to form a diamond.

In CVD, the process starts with a thin diamond "seed" that is placed in a chamber. The chamber is filled with gases containing carbon, and under controlled conditions, the gases are heated, causing them to break down. The carbon atoms then deposit onto the seed, causing it to grow layer by layer into a larger diamond.

Steps in the CVD Process

1. **Preparation**: A thin diamond seed is placed in a vacuum chamber.
2. **Gas Introduction**: Gaseous carbon compounds, usually methane, are introduced into the chamber, along with hydrogen.
3. **Excitation**: A plasma (created by microwaves, lasers, or other means)

excites the gases, causing them to break apart into individual carbon atoms.

4. **Diamond Growth**: The carbon atoms bond to the seed, forming diamond crystals layer by layer.

Diagram of the CVD Process

Here's a simple diagram to show how CVD works:

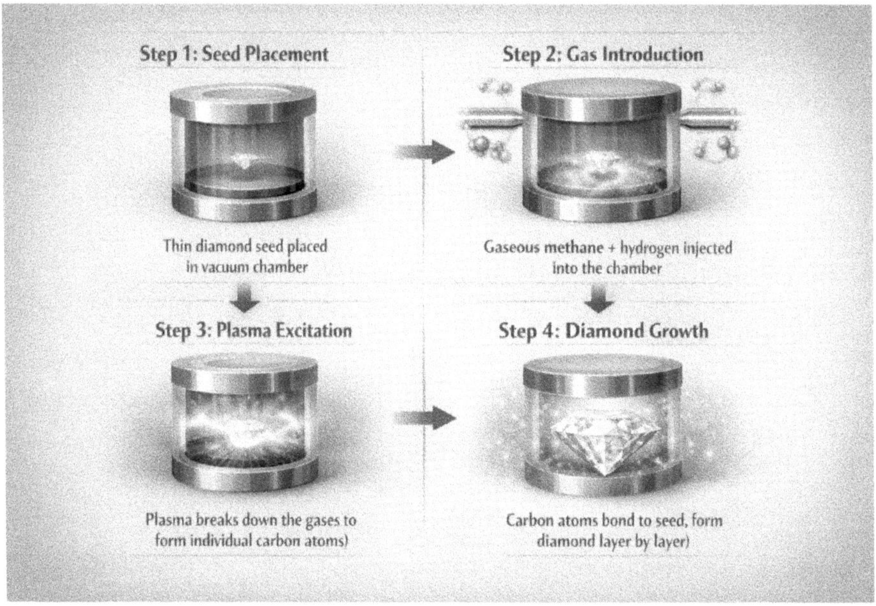

Advantages of CVD:

- **Larger Diamonds**: CVD allows for the production of larger diamonds compared to HPHT, as it is not limited by size constraints.
- **Precision and Control**: The CVD process offers more control over the final product, allowing for greater precision in color, clarity, and overall quality.
- **Environmentally Friendly**: The process uses fewer natural resources than

HPHT and is more energy-efficient in some cases.

Limitations of CVD:

- **Longer Production Time**: While CVD diamonds can grow large, the process takes longer compared to HPHT.
- **Initial Costs**: Setting up the equipment for CVD can be expensive, though the long-term benefits often outweigh the initial investment.

Comparing HPHT and CVD: Which is Better?

Both HPHT and CVD methods have their advantages and limitations, but they share one key characteristic: the diamonds they produce are virtually identical to natural diamonds. The choice between the two methods depends largely on the desired outcome.

- **HPHT** is generally preferred for producing smaller, high-quality diamonds quickly. It also replicates the natural formation process more closely, which can be appealing to consumers who prioritize authenticity.
- **CVD**, on the other hand, is ideal for producing larger diamonds with more precise control over the final product's characteristics. It is particularly beneficial for those seeking larger stones or specific qualities in terms of color and clarity.

Conclusion: A Revolution in Diamond Creation

The ability to create diamonds in a lab is a groundbreaking achievement in material science. Both the HPHT and CVD methods offer distinct advantages that contribute to the growing popularity of lab-grown diamonds. These technologies have made it possible to produce diamonds that are not only stunningly beautiful but also sustainable, ethical, and increasingly accessible to a wider range of consumers.

As we look to the future, the continued evolution of diamond synthesis technologies promises to open new possibilities in the world of diamonds. Whether for use in jewelry, industrial applications, or emerging technologies, lab-grown diamonds are paving the way for a new era in both luxury and sustainability.

Summary

In this chapter, we explored the two primary methods for creating lab-grown diamonds: the High Pressure, High Temperature (HPHT) process and the Chemical Vapor Deposition (CVD) method. We examined how each technique works, their advantages, and their limitations. While HPHT replicates the natural conditions under which diamonds form, CVD provides greater flexibility and control over the final product. Together, these methods are revolutionizing the diamond industry, offering a more sustainable and ethical alternative to traditional mining.

Chapter 4: The Scientific Process Behind Diamond Creation

Lab-grown diamonds are an engineering marvel. The process of creating diamonds in a laboratory, whether through the High Pressure, High Temperature (HPHT) method or the Chemical Vapor Deposition (CVD) method, involves precise scientific techniques that replicate the extreme conditions under which natural diamonds form in the Earth's mantle. These diamonds, though created in controlled settings, are chemically and structurally identical to natural diamonds. To truly appreciate lab-grown diamonds, we must delve into the fascinating science behind their creation.

In this chapter, we'll explore the scientific principles that govern diamond synthesis, from atomic bonding to the growth of the diamond crystal. We'll also examine recent advancements in diamond technology that have made lab-grown diamonds more accessible, cost-effective, and of higher quality. Finally, we'll look at how lab-grown diamonds are transforming from a niche product into a mainstream luxury item.

The Chemistry and Physics of Diamonds

Diamonds are made of carbon atoms that are bonded in a specific, highly ordered arrangement called a crystal lattice. This unique bonding structure gives diamonds their exceptional hardness and brilliance. Understanding how carbon atoms arrange themselves into this lattice structure is essential to understanding both the formation of natural diamonds and the process of growing them in a lab.

At the atomic level, the key to a diamond's structure is its strong covalent bonds. In these bonds, carbon atoms share electrons in a way that creates a stable, repeating pattern. Each carbon atom is bonded to four other carbon atoms in a tetrahedral arrangement, creating a three-dimensional structure. This arrangement of atoms makes diamonds incredibly hard — the hardest known material on Earth. The strength of these bonds is what gives diamonds their exceptional properties, such as their hardness, transparency, and ability to refract light in brilliant ways.

When creating a diamond, either through the HPHT or CVD process, we essentially mimic the conditions that allow carbon atoms to bond in this crystal lattice.

Diagram: Diamond Crystal Structure

Here is a basic illustration of a diamond's crystal structure at the atomic level:

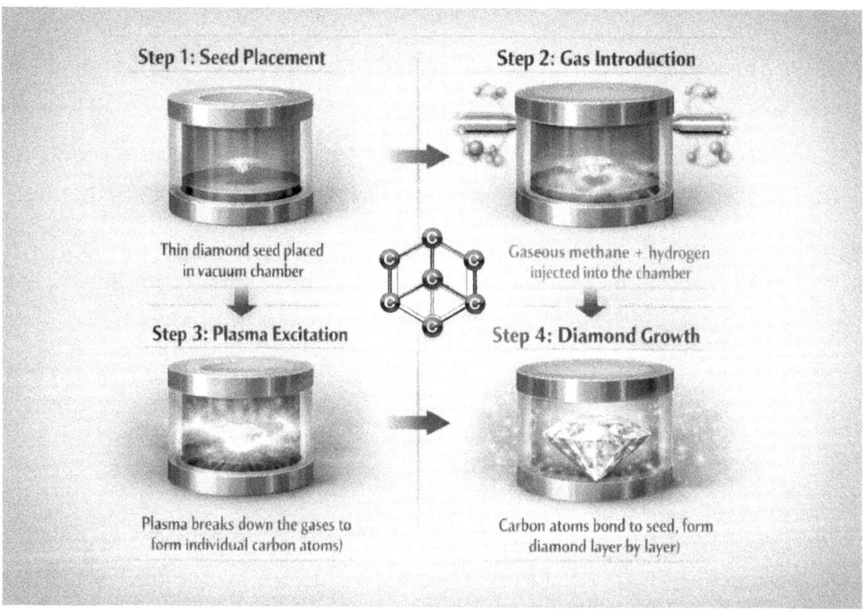

This tetrahedral arrangement, where each carbon atom is bonded to four others, repeats in all directions, forming the lattice structure of a diamond.

How Diamonds Are Created: The Processes Explained

Now that we've covered the basic chemistry of diamond formation, let's take a closer look at how the two primary lab-grown diamond techniques — HPHT and CVD — recreate these extreme conditions in a laboratory environment.

High Pressure, High Temperature (HPHT)

As we discussed in Chapter 1, the HPHT method is designed to replicate the natural conditions under which diamonds form in the Earth's mantle. This method requires high pressure (about 5 gigapascals, or 50,000 times the pressure of Earth's atmosphere) and high temperatures (around 1,300°C to 1,600°C, or 2,400°F to 2,900°F). The carbon source is usually graphite, and the process involves placing a small diamond seed into a metal chamber along with the graphite. Under these extreme conditions, the carbon atoms bond to the seed, forming a diamond crystal.

The key to the HPHT process is creating the right balance of temperature and pressure so that carbon atoms from the graphite source bond in the precise structure needed to create a diamond. The carbon atoms align themselves in the characteristic tetrahedral lattice arrangement, growing the diamond layer by layer.

Chemical Vapor Deposition (CVD)

CVD, developed later as a more versatile alternative to HPHT, operates on a different principle. Instead of mimicking the natural high-pressure environment of the Earth's mantle, CVD uses gases containing carbon, such as methane, which is decomposed under heat and plasma to deposit carbon atoms onto a diamond seed. This process happens in a vacuum chamber that allows for precise control over the temperature, gas mixture, and other factors.

The carbon atoms deposit onto the diamond seed and slowly grow into a full diamond. Over time, the diamond grows larger as more and more carbon atoms bond with the seed. The beauty of CVD lies in its ability to grow large diamonds of high quality and with precision over aspects like color, clarity, and size. It's a slower process than HPHT but offers more control over the final product.

Diagram: Overview of the CVD Process

Here's a simple diagram to illustrate the basic steps of the CVD process:

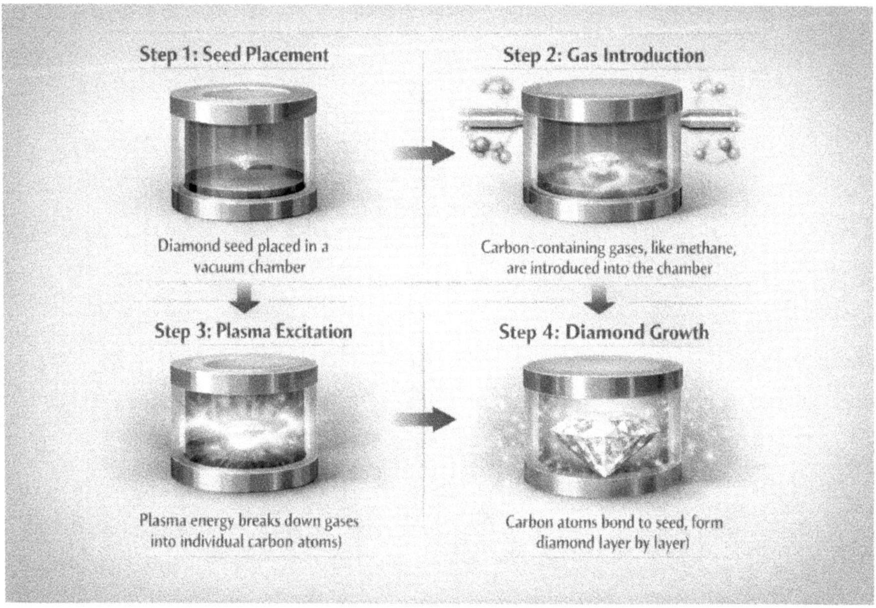

Advances in Diamond Technology

While both HPHT and CVD processes are highly effective at producing lab-grown diamonds, recent advancements in technology have made these methods more efficient, faster, and cost-effective. In particular, improvements in the precision of the equipment used for both methods have allowed for better control over the conditions in which diamonds grow, leading to higher quality stones.

Advances in HPHT Technology

Recent improvements in the HPHT process have focused on increasing the size and quality of diamonds. As technology has advanced, we are now able to produce larger diamonds that meet the high standards expected in the luxury jewelry market. Additionally, the development of new metal alloys in the HPHT chamber has made the process more efficient, reducing the time needed for diamond formation and increasing overall yield.

Advances in CVD Technology

CVD technology has seen rapid improvements in the last decade. Early CVD diamonds were often plagued by impurities and imperfections, but innovations in the control of gas mixtures and plasma energy have greatly improved the quality of CVD diamonds. Today's CVD diamonds can rival mined diamonds in clarity, color, and size. Furthermore, CVD technology has become more scalable, enabling the production of diamonds in various sizes to meet growing market demand.

The Growing Demand for Lab-Grown Diamonds

The evolution of diamond synthesis technology has made lab-grown diamonds increasingly attractive to consumers. The ability to produce diamonds that are virtually identical to natural diamonds, but with a smaller environmental footprint, has led to rising consumer interest in lab-grown options. The key advantages driving this demand include:

- **Sustainability**: Lab-grown diamonds have a significantly lower environmental impact compared to mined diamonds, which require extensive mining operations that can cause significant environmental damage.
- **Ethics**: With concerns about conflict diamonds and unethical mining practices, many consumers are choosing lab-grown diamonds as a more ethically sourced alternative.
- **Affordability**: Lab-grown diamonds are generally more affordable than mined diamonds, making them a more accessible option for a wider range of consumers.

As the technology continues to improve, lab-grown diamonds are likely to become more widespread, and their role in the diamond industry will continue to grow.

Conclusion: The Intersection of Science and Luxury

Lab-grown diamonds represent the fascinating intersection of cutting-edge science and timeless luxury. By replicating the extreme conditions found deep within the Earth, scientists have been able to unlock the secret to diamond creation. The result is a product that is virtually indistinguishable from natural diamonds but comes with significant ethical and environmental benefits.

The continued advancements in diamond technology will likely transform the market even further, with lab-grown diamonds becoming a more common choice for jewelry consumers. The future of diamonds is not just about luxury; it's about creating a sustainable, ethical, and accessible future for one of the world's most treasured gemstones.

Summary

In this chapter, we explored the scientific principles behind diamond creation. We discussed the chemistry of diamonds, how carbon atoms bond in the crystal lattice structure, and how the HPHT and CVD methods replicate the conditions necessary for diamond formation. We also examined recent advancements in diamond technology, which have made lab-grown diamonds more cost-effective, sustainable, and high-quality. As these technologies evolve, lab-grown diamonds are becoming an increasingly popular choice for consumers, offering both luxury and ethical advantages.

Chapter 5: Lab-Grown Diamonds in the Jewelry Industry

The jewelry industry is undergoing a major transformation. As consumers become more environmentally conscious and socially aware, they are shifting their preferences towards products that align with their values. Lab-grown diamonds, which were once a niche product, are now making a significant impact in the jewelry market. From engagement rings to necklaces, lab-grown diamonds are finding their place in the luxury sector, offering consumers a choice that is both ethical and sustainable without compromising on quality.

In this chapter, we will explore the growing popularity of lab-grown diamonds in the jewelry industry. We will discuss the current market trends, how lab-grown diamonds are being used in fine jewelry, and the reasons behind the shift in consumer preferences. We will also examine how luxury brands are integrating lab-grown diamonds into their collections and the impact of this change on traditional diamond suppliers.

The Rise of Lab-Grown Diamonds in the Jewelry Market

The jewelry market has long been dominated by mined diamonds, which have held a special place in the hearts of consumers due to their rarity, beauty, and symbolism. However, in recent years, the demand for lab-grown diamonds has surged, fueled by a number of factors.

Consumer Demand for Ethical and Sustainable Products

The growing awareness about the environmental and social impact of diamond mining has played a major role in the rise of lab-grown diamonds. Traditional diamond mining, particularly in areas with less regulation, has been linked to deforestation, habitat destruction, and unethical labor practices, including child labor. Many consumers, especially younger generations, are looking for alternatives that align with their values. Lab-grown diamonds offer a more sustainable and ethical option, as they are created in a controlled, transparent environment with minimal environmental impact and fair labor practices.

In addition to being ethical, lab-grown diamonds are also seen as a more sustainable choice. Mining for natural diamonds is an energy-intensive process, requiring significant resources and causing environmental degradation. Lab-grown diamonds, on the other hand, can be produced with far less energy and fewer raw materials, making them an eco-friendlier option.

Affordability and Value

Another factor driving the popularity of lab-grown diamonds is their affordability. Lab-grown diamonds typically cost 30% to 40% less than their mined counterparts, allowing consumers to purchase larger or higher-quality diamonds for the same price. This has made lab-grown diamonds an attractive option for those looking to maximize their budget while still investing in a high-quality, beautiful diamond.

As lab-grown diamonds become more mainstream, their prices are expected to continue to drop due to advancements in production technology and economies of scale. This makes them an appealing choice for a growing number of consumers who are looking for luxury at a more accessible price point.

How Lab-Grown Diamonds Are Being Used in Fine Jewelry

Lab-grown diamonds are making their way into all aspects of fine jewelry, from engagement rings to bracelets, earrings, and pendants. The versatility and beauty of these diamonds make them an ideal choice for a wide range of jewelry designs.

Engagement Rings: A Symbol of Love with a Conscience

Engagement rings have traditionally been the cornerstone of the diamond jewelry market. The desire for an engagement ring with a beautiful, high-quality diamond has driven much of the demand for natural diamonds. However, lab-grown diamonds are now becoming a popular alternative for couples who want a diamond that is both stunning and ethically sourced.

One of the main reasons lab-grown diamonds are gaining popularity in engagement rings is their ability to offer consumers the option to purchase a larger or higher-quality diamond within their budget. Since lab-grown diamonds are more affordable, buyers can choose larger stones or higher-quality cuts and clarity without breaking the bank. This allows couples to invest in a more beautiful or meaningful symbol of their commitment.

Other Jewelry Pieces: Necklaces, Earrings, and Bracelets

In addition to engagement rings, lab-grown diamonds are being used in a variety of other jewelry items, including necklaces, earrings, bracelets, and watches. The trend toward more sustainable and ethical products is spreading across the entire jewelry industry, with designers and consumers alike embracing lab-grown diamonds for their versatility and aesthetic appeal.

Jewelry designers are increasingly incorporating lab-grown diamonds into their collections, offering customers the option to wear stunning, eco-friendly diamonds in every piece of jewelry they own. From simple solitaire pendants to elaborate diamond-encrusted bracelets, lab-grown diamonds are being used to create beautiful pieces that are both luxurious and socially responsible.

Customization: Creating Unique Pieces

Lab-grown diamonds also allow for greater flexibility in custom jewelry design. Consumers who want a unique piece that reflects their personal style and values can work with jewelers to create custom designs featuring lab-grown diamonds. Whether it's an engagement ring, a family heirloom, or a special gift, the ability to customize jewelry with lab-grown diamonds allows for a more personalized and meaningful experience.

Consumer Preferences: Shifting Attitudes Toward Lab-Grown Diamonds

The shift in consumer preferences from mined diamonds to lab-grown diamonds is driven by several key factors, including ethical concerns, environmental awareness, and the desire for value.

Ethical Considerations

For many consumers, the appeal of lab-grown diamonds lies in their ethical advantages. Natural diamond mining has long been associated with human rights abuses, including unsafe working conditions, child labor, and exploitation. Lab-grown diamonds, on the other hand, are produced in a controlled environment where labor practices can be monitored and regulated, ensuring fair treatment of workers.

Additionally, lab-grown diamonds do not contribute to the environmental destruction associated with mining. Mining operations can have devastating effects on local ecosystems, including deforestation, water pollution, and habitat destruction. By choosing lab-grown diamonds, consumers are making a conscious decision to avoid supporting these practices.

Environmental Impact

The environmental impact of diamond mining is another key concern for consumers. Traditional diamond mining is resource-intensive, requiring large amounts of water, energy, and raw materials. The process also produces significant amounts of waste, including tailings and toxic runoff that can contaminate local water supplies.

Lab-grown diamonds, by contrast, have a much smaller environmental footprint. While the production of lab-grown diamonds still requires energy, the process is far less destructive than traditional mining. As the technology behind lab-grown diamonds continues to improve, their environmental impact is expected to decrease further, making them an even more sustainable option for consumers.

Price and Value

Affordability is also a major factor driving the shift toward lab-grown diamonds. Lab-grown diamonds typically cost 30% to 40% less than mined diamonds, making them a more budget-friendly option for consumers who still want the beauty and luxury of a diamond. This allows consumers to either purchase a larger diamond or invest in a higher-quality diamond at the same price point.

The lower cost of lab-grown diamonds also makes them more accessible to a wider range of consumers. This democratization of luxury has contributed to the growing demand for lab-grown diamonds, as people from all walks of life are now able to afford beautiful diamond jewelry without the high price tag.

Luxury Brands and Lab-Grown Diamonds: A New Era of Luxury

As lab-grown diamonds gain traction in the market, luxury brands are beginning to take notice. Initially, many high-end jewelry brands hesitated to embrace lab-grown diamonds, preferring to stick with the tradition of natural diamonds. However, as consumer demand for sustainable and ethical products has grown, luxury brands have increasingly incorporated lab-grown diamonds into their collections.

Luxury Brands Integrating Lab-Grown Diamonds

Leading jewelers such as Tiffany & Co., De Beers, and Swarovski have begun offering collections featuring lab-grown diamonds, catering to the growing demand for eco-friendly and ethical luxury. These brands have recognized that lab-grown diamonds are not only a smart business decision but also a way to connect with the younger, more socially conscious consumers who are shaping the future of the luxury market.

For example, De Beers launched its "Lightbox" collection, which features lab-grown diamonds at a lower price point than traditional mined diamonds. This collection allows consumers to enjoy the beauty of diamonds without the associated ethical concerns, providing a more accessible option in the luxury market.

The Impact on Traditional Diamond Suppliers

The rise of lab-grown diamonds is challenging traditional diamond suppliers who rely on mined diamonds. While many established jewelers have adapted by offering both natural and lab-grown options, others are more resistant to change. The increasing popularity of lab-grown diamonds, however, has forced the industry to reconsider the role of mined diamonds in the future of luxury jewelry.

As consumers continue to demand more sustainable and ethically sourced products, traditional diamond suppliers will need to adapt or risk losing market share to lab-grown diamonds. The shift toward lab-grown diamonds is not just a trend but a long-term change in consumer preferences that is reshaping the entire diamond industry.

Conclusion: A New Age of Ethical Luxury

The rise of lab-grown diamonds represents a new era in the jewelry industry. As consumers become more aware of the ethical and environmental issues associated with mined diamonds, they are increasingly turning to lab-grown diamonds as a more sustainable and socially responsible alternative. Lab-grown diamonds are no longer just a niche product but are quickly becoming a mainstream choice for engagement rings, fine jewelry, and luxury items.

With the combination of affordability, ethical sourcing, and environmental sustainability, lab-grown diamonds are reshaping the future of the jewelry market. As the technology continues to improve, lab-grown diamonds will likely become an even more popular choice, offering consumers a way to enjoy the beauty and luxury of diamonds without compromising their values.

Summary

In this chapter, we explored the growing role of lab-grown diamonds in the jewelry industry. We discussed the key factors driving their rise, including ethical concerns, environmental impact, and affordability. Lab-grown diamonds are now being used in a variety of fine jewelry, including engagement rings, necklaces, and bracelets. We also examined how luxury brands are embracing lab-grown diamonds and the impact this shift is having on traditional diamond suppliers. As consumer demand for sustainable and ethical products continues to grow, lab-grown diamonds are poised to play a significant role in the future of the jewelry market.

Chapter 6: Ethical, Environmental, and Social Impact

Lab-grown diamonds are not only a revolution in terms of their technological and economic implications but also represent a profound shift in the jewelry industry when it comes to ethics, sustainability, and social responsibility. As more consumers prioritize sustainability, transparency, and social good in their purchasing decisions, lab-grown diamonds have risen as a strong alternative to mined diamonds. These diamonds are not only beautiful but also carry a different set of values that speak to a growing demand for products that respect the environment, adhere to ethical standards, and contribute positively to society.

In this chapter, we will explore the various ethical, environmental, and social impacts of lab-grown diamonds. We will discuss the environmental benefits of lab-grown diamonds, how they address ethical concerns related to traditional diamond mining, and the broader societal benefits of choosing sustainable and ethically sourced luxury goods.

Environmental Benefits of Lab-Grown Diamonds

Traditional diamond mining is a resource-intensive process that requires the extraction of massive amounts of earth to find just a few carats of diamond. This process not only consumes valuable resources but also causes widespread environmental damage. From deforestation to water pollution, the environmental costs of mining natural diamonds are significant. However, lab-grown diamonds offer a more sustainable alternative.

Reduced Environmental Footprint

The process of creating lab-grown diamonds requires fewer resources than traditional diamond mining. Mining operations are often associated with massive land disturbances, including deforestation, soil erosion, and loss of biodiversity. Additionally, diamond mining uses large amounts of water, and the chemicals involved can contaminate nearby water sources.

In contrast, the creation of lab-grown diamonds, whether through the HPHT or CVD method, is much less disruptive to the environment. The energy consumption required for lab-grown diamonds is significantly lower than that for traditional mining operations. Additionally, lab-grown diamond production can be conducted in controlled environments, allowing for better waste management and reduced pollution.

Lower Carbon Emissions

While both diamond mining and lab-grown diamond production involve energy use, the carbon emissions associated with mining are considerably higher. Mining involves heavy machinery, transport, and processing, all of which contribute to significant greenhouse gas emissions. In comparison, lab-grown diamonds, especially those produced using renewable energy sources, can have a much lower carbon footprint. As more companies adopt sustainable energy practices for lab-grown diamond production, this impact will continue to decrease.

For example, some leading producers of lab-grown diamonds have committed to using solar or hydroelectric power to fuel their operations, further reducing the environmental impact of their products.

Water Conservation

Water usage is another environmental concern tied to traditional diamond mining. The extraction of diamonds from the earth requires large amounts of water, which is often contaminated with chemicals and used in the processing of mined diamonds. Lab-grown diamond production, however, typically uses far less water than the mining process, and the water used in the process can be treated and recycled, further minimizing waste.

Ethical Advantages of Lab-Grown Diamonds

Beyond their environmental impact, lab-grown diamonds are also seen as a more ethical choice compared to mined diamonds. The traditional diamond industry has faced significant criticism over the years for its labor practices and the human rights abuses associated with diamond mining. These issues have prompted calls for greater transparency in the diamond supply chain, and lab-grown diamonds offer a viable solution.

Addressing Conflict Diamonds

One of the most significant ethical concerns surrounding the diamond industry is the issue of "conflict diamonds," also known as "blood diamonds." These diamonds are mined in war zones and sold to fund armed conflict, often with the exploitation of local workers, including child labor. Despite efforts to reduce the trade in conflict diamonds, such as the Kimberley Process Certification Scheme, the issue persists in certain regions.

Lab-grown diamonds, on the other hand, are created in a controlled, transparent environment, ensuring that no human rights abuses are involved in their production. Consumers who purchase lab-grown diamonds can be confident that their purchase does not contribute to the funding of violent conflicts or the exploitation of workers.

Fair Labor Practices

The working conditions in diamond mines can often be hazardous, and workers are frequently subjected to poor conditions and low wages. In some areas, diamond mining also involves child labor, putting young children at risk of injury or exploitation. Lab-grown diamond producers, however, can ensure fair labor practices by regulating working conditions in their controlled environments. This means that no child labor is involved in the creation of lab-grown diamonds, and workers are paid fair wages and work in safe conditions.

In addition, the transparency of the lab-grown diamond supply chain means that consumers can easily trace the origin of their diamonds and confirm that they have been produced in an ethical manner. Many producers also adhere to strict labor standards, ensuring that all workers involved in the process are treated with respect and dignity.

Transparency in the Supply Chain

One of the most significant advantages of lab-grown diamonds from an ethical standpoint is the transparency in the supply chain. Unlike mined diamonds, which can often be difficult to trace back to their exact source, lab-grown diamonds can be traced through every step of the production process. This level of transparency provides consumers with the peace of mind that their purchase aligns with their ethical values.

Many lab-grown diamond producers go further to ensure transparency by offering certification and detailed records of the production process. This can include information on the source of the raw materials, the working conditions, and the energy used to produce the diamonds.

Social Impact: Lab-Grown Diamonds and the Greater Good

Lab-grown diamonds are not only changing the jewelry industry but also contributing to positive social change. Their production has the potential to create new economic opportunities, promote ethical labor practices, and support community development.

Promoting Sustainable Development

In areas where traditional diamond mining has caused social and economic challenges, the growth of the lab-grown diamond industry can offer new opportunities for local communities. Lab-grown diamond production can be established in regions with access to renewable energy sources, creating jobs and economic growth in communities that may not have access to mining resources.

Moreover, because lab-grown diamonds require less land disruption and do not rely on large-scale mining operations, they offer an opportunity for more sustainable economic development. This can help address issues such as land degradation, poor working conditions, and lack of access to education that are often associated with mining industries.

Creating Jobs and Empowering Communities

While the traditional diamond mining industry has been criticized for exploitative labor practices, the lab-grown diamond industry has the potential to create fair-paying jobs in more equitable environments. By focusing on transparency and ethics, the lab-grown diamond sector can empower workers in developing countries and local communities by providing them with safe, well-compensated jobs.

Many lab-grown diamond companies also engage in social responsibility initiatives, contributing to education, healthcare, and community development projects in areas where they operate. This contrasts sharply with the negative impact that diamond mining has often had on local communities, particularly in conflict zones.

Encouraging Consumer Education and Social Responsibility

The rise of lab-grown diamonds is also prompting greater consumer education and awareness about the ethical and environmental implications of their purchasing decisions. As consumers become more informed about the issues surrounding diamond mining, they are increasingly choosing lab-grown diamonds as a way to make more socially responsible purchases.

This shift in consumer behavior is driving other industries to rethink their practices and adopt more sustainable and ethical production methods. The impact of lab-grown diamonds is therefore not limited to the jewelry industry but extends to broader societal changes that emphasize sustainability, fairness, and transparency.

Conclusion: The Path to a More Ethical Future

Lab-grown diamonds are not just a technological innovation; they represent a movement toward more ethical, sustainable, and socially responsible luxury. The environmental benefits of lab-grown diamonds, including reduced carbon emissions, water conservation, and less land disruption, make them a more sustainable choice compared to mined diamonds. Furthermore, lab-grown diamonds address the significant ethical concerns related to conflict diamonds, child labor, and exploitation in the traditional diamond industry.

The social impact of lab-grown diamonds extends beyond the jewelry market, promoting sustainable development, empowering communities, and encouraging consumer education. As more consumers choose lab-grown diamonds, they contribute to a larger shift in the global economy toward products that respect both people and the planet.

Summary

In this chapter, we explored the ethical, environmental, and social impacts of lab-grown diamonds. We discussed how lab-grown diamonds offer a more sustainable and ethical alternative to mined diamonds, addressing concerns such as environmental degradation, conflict diamonds, and labor exploitation. Additionally, we examined the broader societal benefits of lab-grown diamonds, including promoting sustainable development, creating jobs, and encouraging consumer education and social responsibility. As demand for ethically sourced and eco-friendly products continues to rise, lab-grown diamonds are positioned to play a key role in shaping a more responsible and sustainable future in the luxury industry.

Chapter 7: Pricing and Economic Viability

One of the most compelling reasons for the growing popularity of lab-grown diamonds is their affordability. While diamonds have traditionally been associated with exclusivity and high prices, lab-grown diamonds offer a more accessible alternative without compromising on quality or beauty. As the market for lab-grown diamonds continues to evolve, consumers are increasingly drawn to their combination of luxury and value.

In this chapter, we will explore the pricing dynamics of lab-grown diamonds, comparing them to mined diamonds in terms of cost-effectiveness. We will examine the factors that influence the pricing of lab-grown diamonds and discuss how the industry is achieving economic viability. Additionally, we will look at the long-term prospects for the pricing of lab-grown diamonds and how market forces will continue to shape their value in the coming years.

The Price Difference: Lab-Grown vs. Mined Diamonds

The most immediate and noticeable advantage of lab-grown diamonds is their lower cost. On average, lab-grown diamonds can cost 30% to 40% less than their mined counterparts. This price difference is primarily due to the cost of production and the absence of the complex and resource-heavy process of mining. While mined diamonds require extensive labor, land excavation, and energy use, lab-grown diamonds are produced in controlled laboratory environments, where the process is more efficient and requires fewer resources.

How Lab-Grown Diamonds Are Priced

Lab-grown diamonds are priced based on many of the same criteria as natural diamonds, including the 4Cs: cut, color, clarity, and carat weight. However, because lab-grown diamonds are typically less expensive to produce, their retail prices reflect this difference. Let's break down the factors that influence the pricing of lab-grown diamonds.

The pricing of a lab-grown diamond follows the same 4C grading system as natural diamonds, and many lab-grown diamonds are graded by independent gemological laboratories, such as the Gemological Institute of America (GIA), to ensure their quality. The key distinction is that lab-grown diamonds are much less expensive to produce, which allows retailers to offer them at a lower price point.

Cut, Color, Clarity, and Carat (4Cs)

Like natural diamonds, lab-grown diamonds are evaluated based on the 4Cs: cut, color, clarity, and carat weight. These factors determine the quality and price of the diamond.

- **Cut**: The cut of a diamond refers to how well it has been shaped and faceted. Lab-grown diamonds can have precise, symmetrical cuts that contribute to their brilliance and sparkle.
- **Color**: Diamonds are graded on a scale from D (colorless) to Z (light yellow or brown). Lab-grown diamonds can be produced in a range of colors, with many high-quality lab-grown diamonds being virtually colorless.
- **Clarity**: Clarity refers to the presence of internal or external imperfections. Lab-grown diamonds can be created with fewer inclusions, resulting in higher clarity and a more flawless appearance.
- **Carat Weight**: Carat weight measures the size of the diamond. Lab-grown diamonds are available in a variety of carat weights, and because they are generally more affordable, consumers can choose larger stones for the same price as smaller natural diamonds.

Table: Comparison of Pricing Factors for Lab-Grown and Mined Diamonds

Price Range of Lab-Grown Diamonds

The price of a lab-grown diamond varies widely based on the 4Cs, just like mined diamonds. However, due to the lower cost of production, the price of lab-grown diamonds is typically lower for diamonds of similar size, cut, and clarity. For example:

- A 1-carat lab-grown diamond can cost anywhere between $1,000 and $4,000, depending on its quality.
- A similar 1-carat mined diamond could cost between $4,000 and $8,000 or more, depending on its characteristics.

This price difference allows consumers to purchase a larger or higher-quality lab-grown diamond for the same budget, making them a popular choice for engagement rings, wedding bands, and other fine jewelry.

Factors Influencing the Pricing of Lab-Grown Diamonds

Several factors impact the pricing of lab-grown diamonds, from production methods to market demand. Understanding these factors can help consumers and industry players navigate the pricing structure and make informed decisions.

Production Costs

The cost of producing lab-grown diamonds is significantly lower than mining them. The two main methods used to create lab-grown diamonds, HPHT and CVD, are both relatively cost-effective compared to the energy, labor, and equipment required for diamond mining. Additionally, because lab-grown diamonds can be produced on demand, there is no need for the massive infrastructure and land disruption involved in mining operations.

Market Demand

The price of lab-grown diamonds can also be influenced by market demand.

As more consumers become aware of lab-grown diamonds and their benefits, demand is expected to rise. This increased demand could help drive down production costs even further, making lab-grown diamonds even more affordable. Additionally, as more jewelers enter the market, competition will likely lead to more competitive pricing.

Technological Advancements

Technological advances in diamond-growing techniques are continuously improving the efficiency and cost-effectiveness of producing lab-grown diamonds. For example, improvements in CVD technology have allowed for the growth of larger, higher-quality diamonds at a faster rate. As production becomes more efficient, prices are likely to decrease, making lab-grown diamonds even more attractive to consumers.

Consumer Perception and Education

Consumer perception of lab-grown diamonds plays a key role in determining their price. While lab-grown diamonds are becoming more widely accepted, there is still some skepticism among consumers who believe that natural diamonds are superior. As consumer education around the benefits of lab-grown diamonds continues to grow, these diamonds are likely to be seen as a more mainstream, desirable option. This shift in consumer attitudes will further drive demand and influence pricing.

The Economic Viability of Lab-Grown Diamonds

The economic viability of lab-grown diamonds is one of the most important factors driving their success in the marketplace. With the ability to produce high-quality diamonds at a lower cost, lab-grown diamonds offer a compelling value proposition to both consumers and producers.

Lower Production Costs Lead to Competitive Pricing

The cost-effectiveness of lab-grown diamond production allows companies to price their products competitively, offering consumers an affordable option without sacrificing quality. As production techniques improve and economies of scale come into play, the price of lab-grown diamonds is expected to continue to fall, making them even more accessible to a broader range of buyers.

Increasing Market Share

As more consumers opt for lab-grown diamonds, producers are finding new ways to improve the quality and variety of their offerings. This growth in market share means that lab-grown diamonds are likely to become an increasingly important part of the jewelry market. Over time, the demand for lab-grown diamonds will drive prices down and solidify their position as a mainstream luxury product.

Investment Potential

Lab-grown diamonds also hold promise as an investment opportunity. While they are generally less expensive than mined diamonds, their potential for appreciation over time is still being explored. The increasing consumer demand for ethical, sustainable, and transparent products could drive the value of lab-grown diamonds up in the future, making them a potentially profitable investment for buyers who are looking to build a collection.

Conclusion: A Bright Future for Lab-Grown Diamonds

Lab-grown diamonds offer an exciting, more affordable alternative to mined diamonds, with many of the same qualities and beauty. As technology continues to improve and consumer awareness increases, the pricing of lab-grown diamonds is expected to become even more competitive, making them an increasingly popular choice for consumers. Their economic viability, combined with ethical and environmental advantages, ensures that lab-grown diamonds are poised to remain a significant part of the jewelry market for years to come.

Summary

In this chapter, we explored the pricing dynamics of lab-grown diamonds and how they compare to mined diamonds in terms of affordability. We discussed the factors that influence the pricing of lab-grown diamonds, including production costs, market demand, and technological advancements. We also looked at the economic viability of lab-grown diamonds and their long-term prospects. As the market continues to evolve, lab-grown diamonds offer consumers a cost-effective and ethical alternative to traditional diamonds, with the potential for continued price reductions and growth in market share.

Chapter 8: Grading and Quality of Lab-Grown Diamonds

One of the most important aspects of diamonds, whether lab-grown or mined, is the grading system that determines their value. The grading process helps consumers understand the quality of a diamond and the factors that influence its price. The four main characteristics used to evaluate diamonds are known as the 4Cs: Cut, Color, Clarity, and Carat weight. These same criteria are used to assess both natural and lab-grown diamonds.

In this chapter, we will explore how lab-grown diamonds are graded, how they compare to mined diamonds in terms of quality, and the advancements in technology that ensure consistency and precision in the production of lab-grown diamonds. Additionally, we will delve into the grading process and certification for lab-grown diamonds, helping you understand how quality is assessed and what it means for consumers.

The 4Cs: How Lab-Grown Diamonds Are Graded

The 4Cs Cut, Color, Clarity, and Carat weight are the standard measures used to evaluate diamonds. These four factors are essential in determining the overall quality and value of a diamond, and they apply equally to both mined and lab-grown diamonds. Let's break down each factor and explore how lab-grown diamonds are graded based on these criteria.

1. Cut: The Most Important C

The cut of a diamond is perhaps the most critical factor in determining its quality. A well-cut diamond will sparkle and reflect light beautifully, while a poorly cut diamond may appear dull, regardless of its other qualities. The cut refers to how well the diamond has been shaped and faceted to maximize its brilliance.

In the case of lab-grown diamonds, the cut can be particularly important. Since lab-grown diamonds are produced in a controlled environment, there is often more precision in the cutting process, which allows for more symmetrical, precise cuts compared to some mined diamonds.

Grading the Cut

The quality of the cut is graded on a scale ranging from Excellent to Poor. The primary factors considered in the cut grade include:

- **Proportions**: The relationship between the diamond's width, height, and depth, which affects how light is reflected and refracted.
- **Symmetry**: How well the facets are aligned and the overall symmetry of the diamond.
- **Polish**: The smoothness of the diamond's surface, which can impact its overall brilliance.

A well-cut lab-grown diamond typically scores higher on symmetry and polish due to the precision of modern diamond-growing technology.

2. Color: Assessing Diamond Hue

Diamonds are graded on a color scale from D (colorless) to Z (light yellow or brown). The less color a diamond has, the more valuable it is considered. Lab-grown diamonds, like mined diamonds, can vary in color, but most high-quality lab-grown diamonds are created to be colorless or near-colorless. The ability to control the conditions during the creation of lab-grown diamonds means that the final product is often of higher color quality compared to some mined diamonds, especially when it comes to eliminating impurities that may cause discoloration.

Color Grading Scale

- **D-F**: Colorless, the highest grade, with no noticeable color.
- **G-J**: Near colorless, which means the diamond may have a slight color visible only when examined under magnification.
- **K-M**: Faint yellow or brown.
- **N-Z**: Light yellow or brown, often visibly noticeable to the naked eye.

Lab-grown diamonds can be produced with greater precision, so it's not uncommon to find high-quality lab-grown diamonds in the D-F range, which are often harder to find in mined diamonds.

3. Clarity: The Absence of Imperfections

Clarity refers to the presence of internal or external imperfections, called inclusions and blemishes, respectively. These imperfections can occur naturally during the diamond formation process or may result from the manufacturing process. The fewer the inclusions or blemishes, the higher the clarity grade.

Lab-grown diamonds typically exhibit fewer inclusions and blemishes than natural diamonds. The controlled environment in which they are created allows for more consistent and precise production, leading to diamonds with better clarity.

Clarity Grading Scale

Clarity is graded on a scale from Flawless (FL) to Included (I), with several grades in between:

- **Flawless (FL)**: No inclusions or blemishes visible under 10x magnification.
- **Internally Flawless (IF)**: No internal inclusions, but may have external blemishes.
- **Very, Very Slightly Included (VVSI)**: Inclusions are difficult to detect under 10x magnification.
- **Very Slightly Included (VSI)**: Inclusions are minor and visible under magnification.
- **Slightly Included (SI)**: Inclusions are noticeable under 10x magnification.
- **Included (I)**: Inclusions are visible to the naked eye.

Lab-grown diamonds are often found to fall into the higher clarity categories such as FL, IF, or VVSI due to the controlled production process that minimizes imperfections.

4. Carat Weight: The Size of the Diamond

Carat weight refers to the size of the diamond. One carat is equivalent to 0.2 grams. Larger diamonds are generally more expensive, but carat weight alone doesn't determine the value of a diamond. A smaller diamond with a higher cut, clarity, and color may be more valuable than a larger, lower-quality diamond.

Because lab-grown diamonds are often more affordable than mined diamonds, it's not uncommon for consumers to opt for larger lab-grown diamonds within their budget. Lab-grown diamonds are available in a variety of carat weights, allowing consumers to select a size that fits their preferences and budget.

How Lab-Grown Diamonds Are Certified

While the grading process for lab-grown diamonds follows the same principles as mined diamonds, certification plays a crucial role in ensuring that a diamond's quality is accurately represented. Certification is provided by independent gemological laboratories, such as the Gemological Institute of America (GIA), the International Gemological Institute (IGI), and the American Gem Society (AGS).

Certification Bodies

- **GIA**: One of the most well-known and respected organizations in the diamond grading industry. GIA certificates provide detailed information on a diamond's cut, color, clarity, and carat weight.
- **IGI**: Another leading grading laboratory that offers certification for lab-grown diamonds. IGI is known for its rigorous and unbiased grading process.
- **AGS**: Known for its stringent grading system, AGS provides certifications that also assess the cut quality in more detail.

These certifications not only provide a clear picture of a diamond's quality but also offer transparency for consumers, helping them make informed purchasing decisions.

Certified Lab-Grown Diamond Example

The example above shows a high-quality lab-grown diamond that is certified by a reputable organization. Such diamonds are produced with high precision, offering consumers a reliable way to assess quality and value.

Advances in Technology and Precision

The production of lab-grown diamonds has advanced significantly over the past few decades, thanks to improvements in the HPHT and CVD methods. These technological advancements have resulted in diamonds with more consistent quality, greater precision in cutting, and fewer inclusions or imperfections.

Precision in Cutting and Symmetry

One of the advantages of lab-grown diamonds is the ability to achieve a higher level of precision in cutting. Modern cutting techniques, often driven by advanced computer technology, allow for perfect symmetry, making lab-grown diamonds stand out in terms of their sparkle and brilliance. The controlled growth environment ensures that the diamonds produced are more likely to have consistent sizes and shapes, making them easier to cut into symmetrical, well-proportioned diamonds.

Improvement in CVD Technology

CVD technology, in particular, has seen tremendous improvements. Early CVD diamonds were often plagued by impurities and irregularities, but recent advances in the process allow for purer, higher-quality diamonds. The ability to fine-tune the growth process means that CVD diamonds can be created with exceptional clarity and color, often rivaling or exceeding the quality of mined diamonds.

Conclusion: The Future of Lab-Grown Diamond Quality

Lab-grown diamonds are rapidly gaining recognition for their beauty and quality. The grading system for lab-grown diamonds mirrors that of mined diamonds, ensuring that consumers can confidently assess their value. Thanks to advancements in technology and the controlled production environment, lab-grown diamonds are often superior in terms of clarity, color, and symmetry. With reputable certification and transparent grading practices, lab-grown diamonds offer consumers a reliable and ethical choice in the luxury market.

Summary

In this chapter, we explored how lab-grown diamonds are graded based on the same 4Cs as mined diamonds: Cut, Color, Clarity, and Carat weight. We discussed the grading scale for each of these factors and how lab-grown diamonds often outperform mined diamonds in terms of clarity, color, and symmetry. Additionally, we covered the certification process, which ensures that the quality of lab-grown diamonds is accurately represented. Advances in diamond-growing technology have further improved the precision and quality of lab-grown diamonds, making them an attractive and reliable option for consumers.

Chapter 9: Broader Applications of Lab-Grown Diamonds

Lab-grown diamonds are no longer just a choice for jewelry. While their role in the fine jewelry market is significant, these diamonds are also finding applications across many other industries, from high-tech electronics to industrial uses and even quantum computing. The unique properties of diamonds such as their hardness, conductivity, and optical clarity to make them ideal for a variety of technological, industrial, and medical applications.

In this chapter, we will explore the broader uses of lab-grown diamonds, including their role in manufacturing, technology, and emerging fields. We will also look at how these diamonds are being used to revolutionize industries outside the traditional jewelry market.

Industrial Uses: Hardness Meets Functionality

Diamonds are the hardest natural material on Earth, making them highly useful in industries that require durable, long-lasting materials. Lab-grown diamonds, with their identical physical and chemical properties to mined diamonds, offer a cost-effective alternative to traditional diamond tools used in manufacturing and industrial applications.

Cutting, Grinding, and Drilling Tools

Lab-grown diamonds are widely used in cutting, grinding, and drilling tools because of their extreme hardness. These diamonds can cut through even the toughest materials, such as metals, ceramics, and concrete. The ability to produce lab-grown diamonds in controlled environments allows manufacturers to produce tools with specific sizes, shapes, and qualities tailored to the needs of the job.

Common tools that use lab-grown diamonds include:

- **Saw blades**: Used in cutting through stone, metal, and concrete.
- **Grinding wheels**: Essential for polishing and shaping hard materials.
- **Drill bits**: Utilized in mining, construction, and oil drilling for their ability to penetrate hard surfaces.

By using lab-grown diamonds in these tools, manufacturers can benefit from reduced costs, as lab-grown diamonds are generally more affordable than mined diamonds. Additionally, these tools often perform better and last longer, reducing the need for frequent replacements.

Abrasives

Diamonds are commonly used in abrasives due to their hardness. Lab-grown diamond abrasives are used in industries like electronics, automotive, and aerospace for polishing, surface finishing, and fine-tuning precision machinery. Because of their affordability and consistency, lab-grown diamonds are ideal for mass production of abrasives that need to perform under high pressure and extreme conditions.

Technological and Medical Uses: Diamonds Beyond Jewelry

While diamonds have been traditionally used in tools and jewelry, their unique properties make them ideal candidates for cutting-edge technology and medical applications. In particular, the optical, electrical, and thermal conductivity of diamonds are invaluable in several high-tech industries.

Electronics and Semiconductors

Diamonds are not only hard, but they also have excellent thermal conductivity, making them ideal for use in electronic components. Lab-grown diamonds are being used in the production of high-performance semiconductors and electronic devices. These devices benefit from the diamond's ability to dissipate heat, making them more efficient and longer-lasting.

For example, lab-grown diamonds are increasingly being used in:

- **High-power electronic devices**: Components like transistors and diodes that require efficient heat dissipation to function properly.
- **Radiation detectors**: Lab-grown diamonds are used in radiation sensors due to their ability to handle high radiation levels without degradation, making them essential in environments like nuclear reactors or medical imaging.
- **LED technology**: Lab-grown diamonds are also being integrated into LED systems for more durable and efficient light sources.

The unique properties of diamonds help to improve the performance, durability, and efficiency of electronic devices, making them increasingly desirable in advanced technology applications.

Medical Devices

The medical industry is also beginning to utilize lab-grown diamonds for their unparalleled properties. Diamonds have biocompatibility, which means they are not rejected by the body, making them ideal for use in medical tools and devices. Additionally, the hardness of diamonds allows them to be used in surgical tools, where precision and durability are crucial.

Lab-grown diamonds are finding their place in the following medical applications:

- **Surgical tools**: Diamond-coated scalpels and surgical instruments benefit from the sharpness and durability of diamonds, allowing for more precise and clean cuts.
- **Dental instruments**: Lab-grown diamonds are used in dental drills for their ability to cut through tough materials like enamel with minimal friction.
- **Biomedical sensors**: Diamonds can be used in sensors that monitor biological conditions, such as glucose levels in diabetes patients, offering a more accurate and long-lasting alternative to other materials.

The use of lab-grown diamonds in these applications is still in its early stages, but as technology advances, their role in the medical field is expected to grow substantially.

Lab-Grown Diamonds in Emerging Technologies

One of the most exciting areas for lab-grown diamonds is their potential use in **quantum computing** and **advanced optical technologies**. Due to their unique electronic and optical properties, diamonds are ideal for use in developing new types of computers and communication systems.

Quantum Computing

Quantum computing relies on quantum bits (qubits) to perform complex calculations. Unlike traditional bits, which are either 0 or 1, qubits can exist in multiple states simultaneously, allowing quantum computers to solve problems much faster than conventional computers. Lab-grown diamonds play a role in quantum computing because of their ability to hold qubits in a stable state, thanks to the properties of nitrogen-vacancy (NV) centers within the diamond lattice.

Lab-grown diamonds can be engineered with precise NV centers, making them ideal for use in quantum computers. The advantages of using diamonds in quantum computing include:

- **Stable Qubits**: Diamonds have the ability to maintain stable qubits at room temperature, which is a significant breakthrough compared to other materials that require extreme cooling.
- **Fast Information Processing**: Quantum computing using diamond-based qubits can lead to faster, more powerful computing systems for industries like cryptography, artificial intelligence, and complex simulations.

Although quantum computing is still in the experimental stage, lab-grown diamonds are expected to be at the forefront of this technological revolution.

Advanced Optical Technologies

Lab-grown diamonds are also making strides in the field of optics. Diamonds have extraordinary optical properties, including the ability to focus and manipulate light in highly efficient ways. This makes lab-grown diamonds ideal for use in advanced lenses, lasers, and other optical devices.

- **Lenses and Imaging Systems**: Lab-grown diamonds are being used to create high-performance lenses for microscopes, telescopes, and cameras, offering greater clarity and precision.
- **Lasers**: Diamonds can be used to generate high-power, high-quality lasers for various industrial and medical applications.
- **High-performance optical components**: Lab-grown diamonds are being used in optical fibers and components that need to function in high-heat or high-radiation environments, such as in aerospace or nuclear industries.

These advancements in optics offer new possibilities for scientific research and industrial applications.

Other Niche Markets for Lab-Grown Diamonds

Lab-grown diamonds are increasingly being explored for use in a variety of niche markets, beyond traditional jewelry and industrial applications. As the technology behind diamond creation continues to improve, new opportunities will continue to arise.

Watches and Accessories

The luxury watch industry is known for its high standards of quality and precision. Lab-grown diamonds are being incorporated into high-end watches and accessories, adding a touch of elegance and sophistication without the ethical and environmental concerns tied to mined diamonds. Consumers can enjoy the luxury of diamonds in watches, cufflinks, and other accessories, all while making a more sustainable choice.

Art and Design

Lab-grown diamonds are also being used in art and design, including sculptures, decorative pieces, and even interactive installations. Artists and designers are exploring the unique qualities of diamonds, such as their light-reflecting properties and hardness, to create stunning pieces that blend luxury with cutting-edge creativity.

Conclusion: A Future Full of Possibilities

Lab-grown diamonds have already proven their value in the jewelry market, but their applications go far beyond that. From industrial tools to high-tech electronics, medical devices, quantum computing, and advanced optics, the versatility of lab-grown diamonds is beginning to shape many industries. As technology continues to evolve, lab-grown diamonds will likely play an even larger role in the future of manufacturing, technology, and medicine.

Their cost-effectiveness, coupled with their exceptional properties, positions lab-grown diamonds as key players in industries where precision, durability, and sustainability are paramount. Whether in quantum computing, medical applications, or cutting-edge optics, lab-grown diamonds are not just transforming the diamond industry but they are helping to pioneer the future of advanced technology and innovation.

Summary

In this chapter, we explored the broader applications of lab-grown diamonds beyond jewelry. We discussed their role in industries such as manufacturing, technology, and medicine, where their unique properties such as hardness, conductivity, and optical clarity to make them invaluable. From cutting tools to quantum computing, lab-grown diamonds are revolutionizing several sectors, offering cost-effective, sustainable, and high-performance solutions. As technology advances, the potential uses for lab-grown diamonds will only continue to expand, positioning them as essential materials in various high-tech and industrial applications.

Chapter 10: Global Business Structure for Lab-Grown Diamonds

As lab-grown diamonds continue to grow in popularity, their production and distribution structure is evolving. The business side of the lab-grown diamond industry is multifaceted, involving numerous players, including producers, wholesalers, retailers, and grading labs. The industry is also being shaped by supply chains, investment strategies, and regional market dynamics. Understanding the structure of the global lab-grown diamond business is key to grasping how the industry functions and where it is headed in the future.

In this chapter, we will explore the global business structure of lab-grown diamonds, from key players in diamond production to distribution and sales channels. We will also discuss the role of grading and certification bodies, as well as how regulations and business strategies are shaping the future of the lab-grown diamond industry.

The Supply Chain: Key Players in the Lab-Grown Diamond Industry

The lab-grown diamond industry has a relatively short supply chain compared to mined diamonds, which require extensive mining, transportation, and processing. In the lab-grown diamond supply chain, the process is more streamlined, focusing on production, cutting, and grading. The key players involved in this chain include producers, wholesalers, retailers, and third-party grading labs.

1. Producers: The Heart of the Lab-Grown Diamond Industry

The first and most crucial part of the lab-grown diamond supply chain is the production of the diamonds themselves. Lab-grown diamond producers use either the High Pressure, High Temperature (HPHT) method or the Chemical Vapor Deposition (CVD) method to create diamonds in a controlled environment. These producers are the foundational players in the industry.

Some well-known producers of lab-grown diamonds include:

- **Pure Grown Diamonds**: A leader in the field, Pure Grown Diamonds produces high-quality diamonds using CVD technology.
- **Diamond Foundry**: This company is a major player in the lab-grown diamond market and is known for producing diamonds with the highest level of purity and clarity.
- **Altr Created Diamonds**: Known for its ethical approach, Altr is one of the pioneers in the production of lab-grown diamonds.

2. Wholesalers and Distributors: Facilitating the Flow of Diamonds

Once lab-grown diamonds are produced, they are often sold to wholesalers who then distribute them to retailers or direct consumers. Wholesalers play a crucial role in connecting producers with the broader market, ensuring that diamonds reach the right buyers.

These wholesalers often focus on larger bulk sales to jewelry manufacturers, retailers, or online platforms. They buy diamonds directly from the producers and distribute them in various sizes and qualities, making them available for different markets and price points.

3. Retailers: Bringing Lab-Grown Diamonds to Consumers

Retailers are the next link in the supply chain. They can range from online platforms to traditional jewelry stores. Some retailers focus exclusively on lab-grown diamonds, while others may offer both lab-grown and mined diamonds.

Popular online retailers such as **Brilliant Earth** and **James Allen** offer consumers a wide variety of lab-grown diamonds, often with the option for customization. Other companies, like **Blue Nile**, provide a mix of both mined and lab-grown diamonds, catering to a broad range of customer preferences.

4. Third-Party Grading Labs: Ensuring Quality and Transparency

Grading labs, such as the **Gemological Institute of America (GIA)**, **International Gemological Institute (IGI)**, and **American Gem Society (AGS)**, play a critical role in the lab-grown diamond industry. These independent organizations assess the quality of diamonds based on the 4Cs: Cut, Color, Clarity, and Carat weight.

While the grading process for lab-grown diamonds is similar to that of mined diamonds, the primary difference lies in the transparency of the production process. Consumers can feel more confident in their purchase knowing that lab-grown diamonds have been graded independently for quality.

Grading also helps establish the value of a diamond in the market, which is essential for setting fair pricing. Certification from a reputable lab, such as GIA or IGI, is often required by retailers to ensure consumer confidence in the product's authenticity and quality.

Manufacturers and Suppliers: Leading Companies in Diamond Synthesis

Lab-grown diamonds are produced by a small group of companies specializing in the high-tech process of diamond synthesis. These companies use cutting-edge technology to create diamonds that are indistinguishable from mined diamonds, both visually and chemically.

Key manufacturers and suppliers in the lab-grown diamond market include:

These companies are pushing the boundaries of diamond technology and production. As competition grows, more producers are entering the market, bringing down costs and making lab-grown diamonds more accessible to consumers.

Distribution Channels: Making Lab-Grown Diamonds Accessible

Distribution channels for lab-grown diamonds are evolving rapidly. Traditionally, diamonds were sold through brick-and-mortar jewelry stores, but today's market includes online platforms, direct-to-consumer sales, and even luxury brands integrating lab-grown diamonds into their collections.

Online Retailers

Online platforms are a major player in the lab-grown diamond market. Companies such as **Brilliant Earth**, **James Allen**, and **Blue Nile** are popular among consumers who prefer to shop from the comfort of their homes. These platforms provide detailed product information, customizable options, and convenient customer service.

Online retailers allow customers to compare diamonds based on the 4Cs, offering greater transparency and making it easier for them to make informed purchasing decisions. Additionally, online retailers often offer lower overhead costs than physical stores, allowing them to sell lab-grown diamonds at more competitive prices.

Physical Jewelry Stores

Despite the rise of online platforms, physical jewelry stores remain a crucial distribution channel. Major retailers like **Tiffany & Co.**, **Zales**, and **Kay Jewelers** have incorporated lab-grown diamonds into their collections. These retailers offer customers the opportunity to see the diamonds in person and consult with experts before making a purchase. Lab-grown diamonds are increasingly being featured alongside mined diamonds in these stores, allowing consumers to choose between the two options.

Luxury Brands

Luxury brands, once hesitant to embrace lab-grown diamonds, are now integrating them into their offerings. For example, **De Beers** launched its **Lightbox** collection, which features affordable lab-grown diamonds for everyday wear. Other luxury houses are following suit, offering high-end lab-grown diamond jewelry for consumers who want both luxury and sustainability.

Investment and Expansion: Business Strategies for Growth

As demand for lab-grown diamonds grows, so does the need for strategic business planning. Companies in the lab-grown diamond industry are focusing on international expansion, research and development (R&D), and building consumer trust to maintain competitive advantages.

International Expansion

As the market for lab-grown diamonds expands beyond the United States, key players are looking to grow their reach in Europe, Asia, and emerging markets. Companies such as **Diamond Foundry** and **Pure Grown Diamonds** are making strides in international markets, where demand for sustainable products is also rising.

In particular, markets like China and India, where diamond consumption is high, present significant opportunities for growth. As these regions place more emphasis on environmental and ethical practices, lab-grown diamonds could become an even more popular choice for consumers.

Research and Development (R&D)

Continued investment in R&D is crucial to improving diamond production techniques. Companies in the lab-grown diamond industry are constantly looking for ways to reduce costs, improve quality, and increase production efficiency. Technological advancements in CVD and HPHT processes, as well as the ability to create diamonds with higher purity and larger sizes, will drive the industry forward.

Branding and Consumer Trust

Building consumer trust is one of the key challenges for the lab-grown diamond industry. Many consumers still associate diamonds with natural origins, so educating the public on the benefits of lab-grown diamonds is essential. Transparency about production methods, ethical sourcing, and certification can help brands differentiate themselves in the competitive market.

Regulatory Landscape: Government Regulations and Certification

The lab-grown diamond industry is still relatively young, and as such, the regulatory landscape is still developing. Government regulations and certification bodies are becoming increasingly important in ensuring that lab-grown diamonds meet industry standards and that consumers can trust the products they are purchasing.

Regulations and Certifications

Lab-grown diamonds are required to meet specific standards depending on the country of sale. For example, in the United States, the Federal Trade Commission (FTC) has established guidelines that distinguish lab-grown diamonds from mined diamonds. While both are diamonds, lab-grown diamonds must be clearly labeled as such.

Additionally, certification from reputable grading labs like **GIA** and **IGI** plays a key role in ensuring that consumers receive accurate information about the diamonds they are buying. These labs offer certifications based on the 4Cs and also verify the authenticity of lab-grown diamonds, giving consumers confidence in their purchase.

Global Market Trends: Insights into Regional Markets

The demand for lab-grown diamonds is growing worldwide, but regional markets vary in terms of adoption and consumer preferences. Understanding these regional dynamics is key to predicting the future trajectory of the industry.

North America

North America, particularly the United States, has been at the forefront of the lab-grown diamond market. Consumers in this region are increasingly seeking more sustainable and ethical alternatives to mined diamonds. Online retailers and luxury brands are capitalizing on this shift, and major cities like New York and Los Angeles are hubs for both retail and manufacturing of lab-grown diamonds.

Europe

In Europe, demand for lab-grown diamonds is also on the rise, particularly in countries like the United Kingdom, France, and Germany. European consumers tend to be more environmentally conscious, which makes lab-grown diamonds a natural fit for the market. Brands are increasingly offering lab-grown diamonds as part of their collections, and online sales are growing steadily.

Asia

Asia, especially China and India, presents a large untapped market for lab-grown diamonds. While the tradition of mined diamonds is strong in these countries, younger generations are more open to sustainable and ethical alternatives. As awareness about the benefits of lab-grown diamonds grows, the market in Asia is expected to expand significantly in the coming years.

Conclusion: The Evolving Lab-Grown Diamond Industry

The global business structure of lab-grown diamonds is evolving rapidly. From producers and wholesalers to retailers and certification bodies, each part of the supply chain plays a crucial role in making lab-grown diamonds accessible and trustworthy to consumers. As the market grows, more players will enter the space, leading to greater innovation, expanded distribution, and more competitive pricing.

With the growing demand for ethical, sustainable products, lab-grown diamonds are poised to reshape the diamond industry. As companies expand globally, invest in R&D, and build consumer trust, the future of the lab-grown diamond industry looks promising, offering a more sustainable and accessible option for consumers around the world.

Summary

In this chapter, we explored the global business structure of lab-grown diamonds. We discussed the key players involved in the supply chain, from producers to retailers, and examined the role of third-party grading labs in ensuring transparency and quality. We also covered the strategies companies use to expand their reach internationally, invest in R&D, and build consumer trust. As the market for lab-grown diamonds continues to grow, understanding the dynamics of the business structure is essential to predicting the future of the industry.

Chapter 11: Sustainability and Corporate Social Responsibility (CSR)

As the world continues to face growing environmental and social challenges, sustainability and corporate social responsibility (CSR) have become increasingly important for businesses across all sectors. The lab-grown diamond industry, which offers a more sustainable and ethical alternative to traditional mined diamonds, is at the forefront of this shift in consumer behavior. The unique benefits of lab-grown diamonds such as their lower environmental impact and ethical production methods have made them a compelling choice for consumers who are mindful of the planet and the people behind the products they purchase.

In this chapter, we will explore the role of sustainability and CSR in the lab-grown diamond industry. We will examine how lab-grown diamond companies are aligning their operations with global sustainability initiatives, promoting ethical sourcing, and contributing to broader social goals. Additionally, we will look at how CSR strategies are shaping the future of the lab-grown diamond market and influencing consumer decisions.

The Sustainability Advantage of Lab-Grown Diamonds

The environmental impact of diamond mining has long been a concern. Traditional diamond mining requires extensive land disruption, water usage, and the extraction of large amounts of earth to retrieve just a few carats of diamonds. The process also generates significant greenhouse gas emissions and can cause long-term damage to ecosystems. In contrast, lab-grown diamonds are produced in controlled environments that minimize the negative environmental effects associated with mining.

Reduced Carbon Footprint

One of the primary advantages of lab-grown diamonds is their lower carbon footprint compared to mined diamonds. The production of lab-grown diamonds, particularly when powered by renewable energy sources, can result in significantly lower carbon emissions. In contrast, diamond mining involves high-energy operations that contribute to environmental degradation, including significant CO_2 emissions.

Lab-grown diamond companies that use renewable energy sources for their production processes, such as wind or solar power, are leading the way in reducing the industry's overall carbon footprint. These sustainable practices align with the growing global push for carbon-neutral operations in all industries.

Water Conservation

Water use in diamond mining is another environmental concern. Mining operations require large amounts of water to separate diamonds from the surrounding rock, and the process often leads to contamination of local water sources. Lab-grown diamond production, on the other hand, uses significantly less water, and the water used can often be treated and recycled, reducing waste.

The ability to control the production environment in lab-grown diamond facilities means that water usage can be carefully managed, making these diamonds a more sustainable choice for consumers concerned about water conservation.

No Land Degradation

Diamond mining often involves the removal of large amounts of earth, leading to soil erosion, deforestation, and habitat destruction. Lab-grown diamonds, produced in a laboratory setting, do not require large-scale excavation or land disruption, making them an environmentally friendly alternative. The ability to create diamonds in controlled, smaller-scale environments helps preserve natural habitats and ecosystems that would otherwise be impacted by mining activities.

Ethical Production and Transparency

Consumers are increasingly concerned about the ethics of the products they buy, and diamonds are no exception. The traditional diamond mining industry has long faced criticism for human rights abuses, including unsafe working conditions, child labor, and exploitation of workers in some regions. The emergence of lab-grown diamonds offers a solution to many of these issues, as they are produced in ethical, controlled environments where labor standards can be monitored and regulated.

No Child Labor

One of the most significant ethical concerns surrounding mined diamonds is the use of child labor in some mining regions. Children are often forced to work in dangerous conditions, with little regard for their safety or well-being. In contrast, lab-grown diamonds are produced in facilities with regulated working conditions, ensuring that no child labor is involved. This ethical assurance is a significant draw for consumers who are concerned about human rights.

Fair Labor Practices

Lab-grown diamond companies are also committed to providing fair wages, safe working conditions, and opportunities for their employees. By controlling the production environment, these companies can ensure that workers are treated fairly and with respect. Many lab-grown diamond producers also adhere to global labor standards, including the International Labour Organization (ILO) conventions, which protect workers' rights and ensure safe working conditions.

In addition to providing fair wages, some companies also offer additional benefits to their employees, such as healthcare, education, and career development opportunities. These practices reflect the growing demand for businesses to take responsibility for the welfare of their employees and contribute positively to the communities in which they operate.

Lab-Grown Diamonds and the Circular Economy

As the world moves towards a more circular economy where products are reused, recycled, and regenerated lab-grown diamonds are a natural fit. The circular economy model promotes the sustainable use of resources, reducing waste and encouraging the recycling of materials. Lab-grown diamonds align with this vision in several ways.

Recycling and Reusing Materials

Unlike mined diamonds, which are extracted from the earth and used once, lab-grown diamonds can be part of a recycling process. Diamonds, whether lab-grown or natural, can be repurposed and re-used, such as when a piece of jewelry is remade with an old diamond. Furthermore, lab-grown diamonds can be created using recycled carbon from other industrial processes, further enhancing their sustainability.

Closed-Loop Production

Many lab-grown diamond producers are adopting closed-loop production systems. This means that they minimize waste by reusing materials and energy within their production processes. For example, excess carbon can be recycled back into the system to create new diamonds. This reduces the need for new raw materials and helps conserve resources.

By focusing on waste reduction, energy efficiency, and material recycling, lab-grown diamonds contribute to a more sustainable, circular economy, one where the environmental impact of manufacturing is minimized, and resources are used responsibly.

Corporate Social Responsibility (CSR) in the Lab-Grown Diamond Industry

As sustainability and ethical practices become more important to consumers, companies in the lab-grown diamond industry are adopting comprehensive CSR strategies to meet the growing demand for transparency and accountability. CSR initiatives focus on environmental sustainability, social responsibility, and ethical business practices, and lab-grown diamond companies are increasingly aligning their operations with these principles.

Environmental Impact Reporting

Many lab-grown diamond companies now provide detailed reports on their environmental impact. These reports outline key metrics such as carbon emissions, water usage, and energy consumption, helping consumers make informed decisions about their purchases. By publicly sharing this information, companies demonstrate their commitment to sustainability and transparency, building trust with environmentally conscious consumers.

Some companies are also working to obtain third-party certifications that verify their sustainability efforts. For example, the **Carbon Trust** certification helps companies demonstrate their commitment to reducing their carbon footprint. Similarly, **Fair Trade** certification for labor practices helps ensure that products are produced under fair and ethical conditions.

Community Engagement and Social Projects

In addition to environmental efforts, lab-grown diamond companies are increasingly involved in community engagement and social projects. Some companies contribute to educational initiatives, healthcare programs, and other local development efforts in the regions where they operate. By supporting these initiatives, lab-grown diamond companies help improve the quality of life for local communities and ensure that their operations contribute positively to society.

These CSR initiatives reflect the growing demand for businesses to take responsibility not just for their environmental impact but also for their social and economic influence.

Marketing and Consumer Education: Building a Sustainable Brand

A key aspect of the lab-grown diamond industry's success is effective marketing and consumer education. As lab-grown diamonds are relatively new to many consumers, educating them about the ethical, environmental, and economic benefits is crucial for driving demand.

Educational Campaigns

Many lab-grown diamond companies have launched educational campaigns to inform consumers about the benefits of choosing lab-grown diamonds. These campaigns often highlight the environmental advantages of lab-grown diamonds, as well as the ethical concerns surrounding mined diamonds. By positioning lab-grown diamonds as a responsible and sustainable choice, companies are appealing to consumers who prioritize sustainability in their purchasing decisions.

Transparency and Trust

Transparency is central to building consumer trust. Lab-grown diamond companies are increasingly providing information about their sourcing, production processes, and labor practices, helping consumers feel confident in their purchases. By being transparent about the steps they take to reduce environmental impact and ensure ethical production, lab-grown diamond companies foster trust and loyalty among consumers.

Conclusion: The Future of Sustainability in the Lab-Grown Diamond Industry

The lab-grown diamond industry is well-positioned to lead the way in sustainable and ethical luxury goods. By offering an alternative to traditional diamond mining, which is often associated with environmental degradation and unethical labor practices, lab-grown diamonds represent a more responsible choice for consumers. Through lower carbon footprints, reduced water usage, and fair labor practices, lab-grown diamonds are setting new standards for sustainability in the jewelry industry.

As more companies adopt CSR initiatives, improve transparency, and invest in sustainable production methods, the lab-grown diamond industry will continue to grow and evolve. With increasing consumer demand for ethical and sustainable products, lab-grown diamonds are poised to become a mainstream choice for consumers seeking luxury without compromise.

Summary

In this chapter, we examined the sustainability and corporate social responsibility (CSR) efforts within the lab-grown diamond industry. We explored how lab-grown diamonds offer a more environmentally friendly and ethically produced alternative to mined diamonds. Through reduced carbon emissions, water conservation, and fair labor practices, lab-grown diamonds align with global sustainability goals. Additionally, we discussed the growing importance of CSR initiatives, consumer education, and transparency in shaping the future of the industry. As sustainability becomes an even greater focus for consumers, lab-grown diamonds will continue to lead the way in creating a more responsible and ethical luxury market.

Chapter 12: The Future of Lab-Grown Diamonds: A New Era for Luxury

As the lab-grown diamond industry continues to gain momentum, its future looks bright. With technological advancements, changing consumer preferences, and a growing commitment to sustainability, lab-grown diamonds are poised to reshape the diamond industry in the coming decades. These diamonds are not only changing the way we think about luxury, but they also offer a sustainable, ethical, and affordable alternative to mined diamonds.

In this chapter, we will explore the future of lab-grown diamonds, examining current trends, future projections, and the challenges and opportunities the industry will face. We will also look at how consumer education, marketing, and technological innovation will drive the continued growth of the industry, making lab-grown diamonds a mainstream choice for luxury goods.

Current Trends and Projections for Lab-Grown Diamonds

The lab-grown diamond industry has seen remarkable growth in recent years, driven by factors such as sustainability, ethical sourcing, and increasing consumer awareness. As the market continues to evolve, several key trends are shaping the future of lab-grown diamonds.

1. Rising Consumer Demand for Ethical and Sustainable Products

One of the driving forces behind the growth of lab-grown diamonds is the increasing demand for sustainable and ethically sourced luxury products.

As more consumers become aware of the environmental and social issues associated with mined diamonds, many are turning to lab-grown diamonds as a more responsible option.

The rise of the eco-conscious consumer has made sustainability a key consideration in purchasing decisions. Lab-grown diamonds offer a clear solution to the ethical concerns surrounding traditional diamond mining, such as conflict diamonds and child labor. As sustainability becomes more important to consumers, the demand for lab-grown diamonds is expected to rise steadily.

2. Technological Advancements in Diamond Production

As technology continues to improve, the production of lab-grown diamonds is becoming faster, more efficient, and more affordable. Both the High Pressure, High Temperature (HPHT) and Chemical Vapor Deposition (CVD) methods have seen significant advancements, resulting in better-quality diamonds and reduced production costs.

Future advancements in diamond-growing technology will likely focus on increasing the size and purity of diamonds, reducing the energy required for production, and improving the overall speed of the manufacturing process. These innovations will make lab-grown diamonds more accessible and affordable for a larger number of consumers.

3. Market Expansion and Global Growth

The lab-grown diamond market is not limited to North America and Europe. As awareness of the benefits of lab-grown diamonds grows, markets in Asia, particularly China and India, are emerging as key growth areas. Both of these regions have large populations and a growing appetite for luxury goods, making them ripe for the expansion of the lab-grown diamond market.

With increased demand for both ethical products and luxury items, companies in the lab-grown diamond industry are expanding their operations into these regions. International expansion is a critical strategy for growth, and many lab-grown diamond producers are now targeting global markets.

As the market continues to expand globally, the lab-grown diamond industry will face new challenges but will also benefit from increased access to new customers and regions.

Consumer Education and Marketing: Key Drivers for the Future

As the market for lab-grown diamonds grows, educating consumers about their value, quality, and benefits will be essential to the industry's continued success. Unlike mined diamonds, lab-grown diamonds are still relatively new to many consumers, and misinformation or confusion about their quality and origins can hinder growth.

1. Shifting Perceptions of Luxury

Traditionally, diamonds have been considered a symbol of rarity, luxury, and exclusivity. However, lab-grown diamonds are challenging this notion by offering a more affordable, sustainable, and ethical option. As the benefits of lab-grown diamonds become more widely understood, luxury will increasingly be defined by the values of sustainability, transparency, and ethical production, rather than rarity.

Lab-grown diamonds, in many ways, embody the future of luxury. They provide consumers with the opportunity to own high-quality, beautiful diamonds that align with their values—without the ethical or environmental concerns associated with mined diamonds. As this shift in consumer perception continues, lab-grown diamonds are likely to become the new standard for luxury.

2. Increasing Transparency in the Diamond Industry

Transparency is key to building consumer trust, and the lab-grown diamond industry is leading the way in this area. With consumers becoming more conscious of the ethical implications of their purchasing decisions, companies in the lab-grown diamond sector are prioritizing transparency about their production methods, sourcing, and labor practices.

Lab-grown diamond producers are increasingly providing consumers with information about their diamonds, including details about the technology used to grow them, their environmental impact, and the ethics behind their production. Third-party certifications and transparent grading systems help ensure that consumers can make informed choices about the diamonds they purchase.

3. Engaging and Educating Consumers

The lab-grown diamond industry is using a variety of marketing strategies to engage and educate consumers. Social media platforms, digital advertising, and educational content are being used to inform consumers about the benefits of lab-grown diamonds and to create awareness about the environmental and ethical issues surrounding traditional diamond mining.

By telling the story of lab-grown diamonds—from the technology behind their creation to the positive impact they have on the environment and society—companies can build stronger connections with consumers and differentiate themselves from traditional diamond retailers.

Challenges and Opportunities Ahead

While the future of lab-grown diamonds looks promising, there are still several challenges and opportunities that the industry will need to navigate in order to sustain its growth.

Challenges

1. Overcoming Consumer Skepticism

Many consumers are still unfamiliar with lab-grown diamonds or may harbor misconceptions about their quality. Overcoming skepticism will require continued education and marketing efforts. Transparency, certification, and strong consumer engagement are critical for building trust and confidence in lab-grown diamonds.

2. Competition from Mined Diamonds

The mined diamond industry, with its long history of luxury and prestige, remains a significant competitor to lab-grown diamonds. Traditional diamond companies may resist the rise of lab-grown diamonds, and it will take time to shift public perception fully. However, with the growing demand for ethical and sustainable products, lab-grown diamonds are likely to continue gaining market share.

Opportunities

1. Expanding Product Offerings

As the market for lab-grown diamonds grows, there is an opportunity to expand product offerings beyond traditional engagement rings and jewelry. Lab-grown diamonds could find their place in watches, accessories, and even in industries such as technology, medicine, and industrial applications. This diversification could open up new markets and revenue streams for companies in the lab-grown diamond industry.

2. Increasing Collaboration with Luxury Brands

Lab-grown diamonds are gaining traction among high-end jewelry brands, with some major names embracing the trend. As luxury brands continue to experiment with lab-grown diamonds, the acceptance of these diamonds as a symbol of luxury is likely to increase. Collaborations with established luxury brands will elevate the status of lab-grown diamonds and solidify their position in the luxury market.

3. Enhanced Sustainability and Ethical Practices

Sustainability and ethics will remain at the forefront of the lab-grown diamond industry's future. As technology advances, the carbon footprint and environmental impact of producing lab-grown diamonds will continue to decrease. The industry has the opportunity to lead by example, demonstrating how luxury can be redefined through sustainability, fair labor practices, and transparency.

Conclusion: A Bright Future for Lab-Grown Diamonds

The future of lab-grown diamonds is marked by increasing consumer demand for sustainability, transparency, and ethical practices. As technology continues to evolve, lab-grown diamonds will become more accessible, affordable, and desirable to a growing number of consumers. The industry is well-positioned to not only revolutionize the diamond sector but to redefine luxury itself by offering a product that is both beautiful and responsible.

Lab-grown diamonds are not just an alternative to mined diamonds—they represent a new era in which luxury and sustainability go hand in hand. The growing global demand for these diamonds is proof that consumers are ready to embrace a more ethical, sustainable, and accessible form of luxury.

Summary

In this chapter, we explored the future of lab-grown diamonds, highlighting key trends and projections for the industry. We discussed the role of technological advancements, the growing demand for sustainable products, and the increasing consumer preference for ethical and transparent brands. Additionally, we examined the challenges and opportunities ahead for the industry, including overcoming consumer skepticism and expanding into new markets. With continued innovation and a focus on sustainability, the lab-grown diamond industry is set to reshape the future of luxury.

Chapter 13: Conclusion: The Transformative Impact of Lab-Grown Diamonds

The emergence of lab-grown diamonds represents more than just a technological advancement—it marks a transformation in the luxury goods industry, offering a more ethical, sustainable, and accessible alternative to mined diamonds. Over the past few decades, the lab-grown diamond industry has grown significantly, driven by advances in production technology, increasing consumer demand for ethical products, and a growing global focus on sustainability. Lab-grown diamonds have become a powerful symbol of how innovation can align with consumer values and reshape an entire industry.

In this final chapter, we will reflect on the key points discussed throughout the book, summarize the transformative impact of lab-grown diamonds on the jewelry and luxury goods market, and look at the long-term implications for both the diamond industry and consumers. We will also explore the future of lab-grown diamonds and their role in shaping the ethical and sustainable luxury market of tomorrow.

Key Takeaways: The Rise of Lab-Grown Diamonds

Lab-grown diamonds have gained considerable traction in recent years, and their rise can be attributed to a combination of factors. Here are some key takeaways that summarize the growing significance of lab-grown diamonds:

1. **Ethical and Sustainable Choice**: Lab-grown diamonds provide a solution to the ethical and environmental concerns surrounding mined diamonds. By eliminating practices like child labor, exploitation, and environmental destruction associated with traditional diamond mining, lab-grown diamonds offer a more responsible alternative for consumers who prioritize sustainability and ethical sourcing.
2. **Technological Innovation**: Advances in production technologies, such as High Pressure, High Temperature (HPHT) and Chemical Vapor Deposition (CVD), have allowed lab-grown diamonds to be produced with exceptional quality and affordability. These innovations have made it possible to produce diamonds that are indistinguishable from their mined counterparts in terms of appearance, physical properties, and value.
3. **Affordability and Accessibility**: Lab-grown diamonds are generally more affordable than mined diamonds, offering consumers the opportunity to own high-quality diamonds at a fraction of the price. This affordability has made lab-grown diamonds a viable option for a broader range of consumers, including those who might not have been able to afford traditional mined diamonds.
4. **Consumer Demand for Transparency and Ethical Practices**: The growing awareness of the negative impacts of diamond mining has shifted consumer preferences towards more sustainable and ethically sourced products. Lab-grown diamonds, produced in controlled environments with traceable supply chains, align with this shift in consumer behavior, providing transparency and assurance to buyers.

The Lab-Grown Diamond Industry: A New Era for Luxury

Lab-grown diamonds are not just a trend; they represent a new era in the luxury goods market. This era is defined by a shift towards sustainable luxury, where the focus is not just on aesthetics and exclusivity but also on the broader impact that products have on the planet and society. Lab-grown diamonds have emerged as the frontrunners in this movement, offering luxury without the ethical and environmental compromises associated with traditional diamond mining.

Rewriting the Story of Luxury

For generations, diamonds have been seen as symbols of wealth, status, and eternal love. Their rarity and the mystique surrounding their natural origins have made them a cherished commodity. However, as consumer values evolve and the importance of ethical and environmental considerations grows, the very definition of luxury is changing.

Lab-grown diamonds challenge the traditional narrative by offering a product that is both luxurious and responsible. As more consumers seek to align their purchases with their values, lab-grown diamonds provide a way to enjoy luxury without contributing to environmental degradation or exploitation. This shift is particularly evident among younger generations, who are more likely to prioritize sustainability, transparency, and ethical production in their purchasing decisions.

The Role of Lab-Grown Diamonds in Redefining Luxury

As the demand for lab-grown diamonds continues to rise, they are reshaping what it means to be "luxurious." Once considered a cheaper alternative, lab-grown diamonds are increasingly seen as a symbol of modern luxury—one that is both beautiful and responsible. This shift reflects broader societal changes, where luxury is no longer solely about rarity but also about the values that a product represents.

The increasing integration of lab-grown diamonds into the collections of luxury brands, along with the growing acceptance of these diamonds in mainstream markets, indicates that we are entering a new era for diamonds. This era is defined not by scarcity, but by sustainability, innovation, and ethical responsibility.

The Growing Impact on the Diamond Industry

The growth of the lab-grown diamond industry has already had a significant impact on the traditional diamond sector. As consumer demand for sustainable and ethically sourced products continues to increase, the mined diamond industry has had to respond by adapting to changing market conditions.

Competition and Market Dynamics

As lab-grown diamonds become more widely accepted, traditional diamond companies are adjusting their strategies. Some companies have embraced lab-grown diamonds, offering them alongside their mined diamonds to appeal to the growing segment of ethically conscious consumers. Others are focusing on the rarity and heritage of mined diamonds to maintain their appeal in the luxury market.

The competition between mined and lab-grown diamonds is expected to intensify as consumer demand shifts. While mined diamonds will always hold a certain appeal due to their natural origins, lab-grown diamonds provide a viable alternative that is aligned with the values of today's consumers. Over time, lab-grown diamonds are likely to capture a larger share of the market, particularly as their production methods improve, costs decrease, and consumer awareness grows.

Consumer Education: A Key to Lab-Grown Diamond Success

For lab-grown diamonds to continue their growth and acceptance, consumer education is crucial. Many consumers are still unfamiliar with lab-grown diamonds or may harbor misconceptions about their quality and value. Clear communication about the benefits of lab-grown diamonds—such as their environmental impact, ethical production, and affordability—is essential to building trust and increasing market share.

Companies in the lab-grown diamond sector are already using various strategies to educate consumers, from informative marketing campaigns to transparency in production and grading. As consumer awareness of the ethical and environmental benefits of lab-grown diamonds continues to increase, the demand for these products will likely grow exponentially.

The Role of Transparency and Certification

Transparency in the production and sourcing of diamonds is one of the key selling points for lab-grown diamonds. Consumers are more likely to trust a product when they have access to clear and verifiable information about its origins. Independent certification from reputable grading labs, such as the **Gemological Institute of America (GIA)** and **International Gemological Institute (IGI)**, is essential to ensuring the quality and authenticity of lab-grown diamonds.

Lab-grown diamond companies that prioritize transparency and provide clear information about their sustainability practices, sourcing methods, and ethical production will gain a competitive edge in the market. This transparency will help to build trust with consumers, making it easier for them to make informed purchasing decisions.

The Path Forward: Challenges and Opportunities

While the lab-grown diamond industry has experienced tremendous growth, there are still challenges ahead. The most significant challenge is overcoming the cultural and emotional attachment that many consumers have to natural diamonds. For generations, diamonds have been symbols of love, status, and rarity, and shifting these perceptions will take time.

However, as more consumers embrace lab-grown diamonds for their ethical and environmental benefits, these challenges will be addressed. The growing trend of sustainable luxury, combined with advancements in technology, will continue to drive the growth of lab-grown diamonds.

Opportunities for Innovation

There are significant opportunities for innovation in the lab-grown diamond industry. Advances in diamond-growing technology will likely result in diamonds of even higher quality, with fewer inclusions and more precise cuts. The potential for customization will also become more prevalent, allowing consumers to design their own diamonds in a way that reflects their personal values and style.

In addition, the expansion of lab-grown diamonds into new markets, such as industrial applications, electronics, and medicine, will further fuel the growth of the industry. As technology advances, lab-grown diamonds will continue to play a pivotal role in industries that rely on the unique properties of diamonds, such as quantum computing, electronics, and high-performance materials.

Conclusion: Lab-Grown Diamonds Shaping the Future of Luxury

Lab-grown diamonds are not just an alternative to mined diamonds—they are reshaping the very definition of luxury. By offering a product that is both beautiful and responsible, lab-grown diamonds are meeting the demand for ethical, sustainable, and affordable luxury. As the industry continues to grow, lab-grown diamonds will increasingly become a mainstream choice for consumers seeking a more sustainable and ethical option.

The future of the lab-grown diamond industry is bright. With technological advancements, increased consumer awareness, and a growing commitment to sustainability, lab-grown diamonds are set to revolutionize the luxury market. As they continue to gain market share, lab-grown diamonds will play a key role in reshaping the diamond industry for generations to come.

Summary

In this concluding chapter, we reflected on the transformative impact of lab-grown diamonds on the jewelry and luxury goods market. We discussed key trends driving the growth of lab-grown diamonds, including consumer demand for ethical and sustainable products, technological advancements in diamond production, and the increasing market acceptance of lab-grown diamonds. The rise of lab-grown diamonds represents a new era in luxury, one where sustainability, innovation, and transparency redefine the meaning of luxury. As the industry continues to grow, the future of lab-grown diamonds is poised to reshape the diamond industry and offer consumers a new, more ethical standard of luxury.

Appendix A: Glossary of Terms

Here is a list of common terms used throughout the book to help clarify key concepts related to lab-grown diamonds and the diamond industry in general.

- **4Cs**: The four criteria used to grade diamonds: Cut, Color, Clarity, and Carat weight. These factors determine a diamond's quality and value.
- **CVD (Chemical Vapor Deposition)**: A method used to grow lab-grown diamonds by introducing gases (like methane) into a vacuum chamber and then using energy to break the gases down, allowing carbon atoms to bond and form diamonds.
- **HPHT (High Pressure, High Temperature)**: A method that mimics the natural diamond-forming process by applying extreme heat and pressure to create diamonds from carbon.
- **Conflict Diamonds**: Diamonds mined in war zones and sold to fund armed conflict, often associated with human rights abuses and exploitation.
- **Carat Weight**: A unit of weight used to measure the size of a diamond. One carat is equivalent to 0.2 grams.
- **Clarity**: A measure of the number and visibility of inclusions (internal flaws) or blemishes (external flaws) in a diamond.
- **Cut**: Refers to how well a diamond has been shaped and faceted, influencing its brilliance and sparkle.
- **Color**: A measure of how colorless a diamond is. The scale ranges from D (colorless) to Z (light yellow or brown).

- **Gemological Institute of America (GIA)**: One of the most respected and recognized diamond grading organizations in the world.
- **Diamond Foundry**: A company that specializes in producing lab-grown diamonds using CVD technology.
- **Carbon Footprint**: The total amount of carbon dioxide emissions produced during the creation or lifecycle of a product.

Appendix B: Lab-Grown Diamond Certification and Grading

Lab-grown diamonds, like mined diamonds, are graded based on the 4Cs. Reputable third-party laboratories provide certification for these diamonds, offering transparency and assurance regarding their quality. Below are the primary certification organizations and the grading process.

Key Certification Organizations:

1. **Gemological Institute of America (GIA)**

- Provides detailed reports on diamonds based on the 4Cs. They grade both mined and lab-grown diamonds using the same standards.
- The GIA is widely considered the gold standard in diamond certification and grading.

1. **International Gemological Institute (IGI)**

- One of the leading diamond grading and certification organizations in the world. IGI provides reports for both lab-grown and mined diamonds, offering an easy-to-understand grading system.
- IGI also certifies the cut, clarity, color, and carat weight of diamonds.

1. **American Gem Society (AGS)**

- Known for its rigorous grading system, AGS focuses heavily on the cut of the diamond, which it considers to be the most important factor in the diamond's overall beauty.

- AGS provides certified grading reports for both lab-grown and mined diamonds.

Diamond Grading Report Example

What is Included in a Diamond Certificate?

1. **Cut**: The quality of the cut is based on symmetry, proportions, and polish. A well-cut diamond will reflect light better and have greater brilliance.
2. **Color**: The absence of color is graded on a scale from D (colorless) to Z (light yellow or brown).
3. **Clarity**: The presence of inclusions or blemishes in the diamond is evaluated under magnification. A diamond with fewer inclusions is generally more valuable.
4. **Carat Weight**: The weight of the diamond is measured in carats, with one carat equaling 0.2 grams.
5. **Diamond Shape**: The shape refers to the diamond's overall form, such as round, princess, emerald, or cushion. This is not part of the grading, but it is often noted on the certificate.
6. **Laser Inscription**: Some lab-grown diamonds feature a laser inscription on the girdle of the diamond, which provides an additional level of identification and authenticity.

Appendix C: Lab-Grown Diamond Production Process – Step-by-Step

The production of lab-grown diamonds involves highly sophisticated techniques. Below is an overview of the two primary methods—HPHT and CVD—used to grow diamonds in a laboratory setting.

1. High Pressure, High Temperature (HPHT) Process

- **Step 1: Carbon Source Selection**A carbon source, typically graphite, is used to create a diamond. A small diamond seed, which will serve as the nucleus for the diamond, is placed in a metal chamber.

- **Step 2: Creating Extreme Conditions**The chamber is subjected to temperatures exceeding 1,300°C and pressures of 5 gigapascals, replicating the conditions found in the Earth's mantle.
- **Step 3: Diamond Formation**Under these extreme conditions, the carbon atoms bond together in the crystal lattice structure that defines a diamond. The diamond grows layer by layer around the seed.
- **Step 4: Cooling and Extraction**After the diamond has reached the desired size, the chamber is slowly cooled, and the diamond is carefully extracted from the metal chamber.

2. Chemical Vapor Deposition (CVD) Process

- **Step 1: Seed Placement**A thin diamond seed is placed in a vacuum chamber. The chamber is filled with gases, typically methane and hydrogen, which will serve as the carbon source.
- **Step 2: Plasma Creation**The gases are excited using plasma energy, usually created by microwaves, lasers, or other forms of energy. This breaks down the gases, allowing carbon atoms to separate and bond with the diamond seed.
- **Step 3: Diamond Growth**As carbon atoms bond to the seed, the diamond crystal grows layer by layer, forming a larger diamond over time.
- **Step 4: Harvesting the Diamond**Once the desired size and quality are reached, the diamond is carefully removed from the chamber.

Appendix D : Environnemental Impact – Comparative Analysis

As the demand for ethical and sustainable products continues to grow, lab-grown diamonds are becoming a popular alternative to mined diamonds. Below is a comparison of the environmental impact of mined versus lab-grown diamonds.

Lab-grown diamonds offer a much lower environmental impact, making them a compelling option for consumers who are concerned about sustainability.

Appendix E: Lab-Grown Diamond Market Overview

The lab-grown diamond industry has experienced significant growth, driven by increasing consumer demand for ethical and sustainable alternatives to mined diamonds. Below is a summary of key market trends and statistics.

Global Lab-Grown Diamond Market Size

- The global market for lab-grown diamonds was valued at approximately **$1.9 billion** in 2021 and is expected to grow at a compound annual growth rate (CAGR) of **15-20%** over the next decade.
- As more consumers seek ethical, affordable, and sustainable options, the market for lab-grown diamonds is poised to surpass **$4.5 billion** by 2030.

Regional Market Insights

- **North America**: North America continues to be the largest market for lab-grown diamonds, driven by high consumer awareness, technological advancements, and the increasing demand for ethical products.
- **Europe**: Europe is also experiencing significant growth, with countries like the UK, Germany, and France leading the charge in adopting lab-grown diamonds.
- **Asia**: With the rapid growth of the middle class in countries like China and India, the Asian market for lab-grown diamonds is expanding quickly, especially in the high-end jewelry market.

Consumer Preferences

- **Sustainability and Ethics**: Consumers are increasingly prioritizing products that align with their values, particularly when it comes to environmental and ethical concerns.
- **Affordability**: Lab-grown diamonds offer a more affordable alternative to mined diamonds, allowing consumers to access larger or higher-quality diamonds within their budget.
- **Customization**: Many consumers are choosing lab-grown diamonds

for their ability to offer more customization options, including bespoke designs and unique diamond shapes.

Appendix F: Resources for Further Reading

For readers interested in learning more about lab-grown diamonds, sustainability in the diamond industry, or the technological advancements in diamond production, the following resources provide valuable information:

- **The Diamond Producers Association**: A global trade organization representing the natural diamond industry. Website: www.diamondproducers.com
- **Gemological Institute of America (GIA)**: A leading organization for diamond certification and education. Website: www.gia.edu
- **International Gemological Institute (IGI)**: Provides certifications for diamonds and gemstones. Website: www.igi.org
- **Sustainable Development Goals (SDGs) by the United Nations**: Learn about the global goals for sustainability and ethical business practices. Website: www.un.org/sustainabledevelopment

These resources offer further insight into the diamond industry's future and its growing emphasis on sustainability and ethics.

www.ingramcontent.com/pod-product-compliance
Lightning Source LLC
Chambersburg PA
CBHW041300210926
51847CB00009B/93
9788846246912